NEW ZEALAND WINE GUIDE

An Introduction to
the Wine Styles & Regions
of New Zealand

CELIA HAY

This book supports the delivery of unit standards currently registered by the New Zealand Qualifications Authority:

22912	Food and Beverage Service Evaluate Wine
4637	Food and Beverage Service Demonstrate Knowledge of New Zealand Wines and Producers
23060	Food and Beverage Service Demonstrate knowledge of viticulture and wine making

I have also drawn material from the Wine and Spirit Education Trust® for whom the New Zealand School of Food and Wine has been an Approved Programme Provider since 2004.

Cover: Dog Point Vineyard, Lower Brancott Valley, Marlborough. Photography by Kevin Judd.

ACKNOWLEDGEMENTS

A big thank you to the Hospitality Training Trust for its generous grant supporting the publication of this book as a resource and textbook for wine education in New Zealand.

Photographs: Kevin Judd, for the use of his wonderful photographs which visually give important context and an essential lift to a lot of text. Therese and Hans Herzog, Millton Estate, Riversun Nursery, Villa Maria, Misha's Vineyard, Auntsfield Estate, Auckland War Memorial Museum, Alexander Turnbull Library.

Maps: Geographx for regional maps; New Zealand Winegrowers for map of New Zealand wine regions; NIWA.

For editorial comment and advice: Joelle Thomson, Lynnette Hudson, Therese Herzog, Kevin Judd, Paul Tudor MW, Michael Brajkovich MW, Tim Hanni MW, New Zealand Winegrowers.

Wine and Spirit Education Trust® WSET

Celia Hay

Every effort has been made to ensure the information included in this book is as accurate as possible. We welcome feedback on ways to improve the content.

Book design: Donna Hoyle Design Ltd

Hay Publishing
PO Box 106750
Auckland

www.nzwinebook.com

© Celia Hay 2014

ISBN 978-0-473-28066-6

CONTENTS

INTRODUCTION

Why study wine?

For me, there are many reasons that make studying about the world of wine such a fascinating subject.

For a start, wine has been made for over 5000 years, yet the winemaking process, in its most basic form, remains unchanged and it is still possible to produce wine to tempt us. I also like to think of this as a personal journey of discovery that not only develops an appreciation of different aromas and wine styles but helps you to learn more about your palate and how things taste to you. If this is not enough, grapes flourish in some of the most beautiful locations on earth, producing wines that convey a unique character and sense of place. This is international travel, in a glass. Wine is also aspirational. The selection of a particular bottle of wine has the power to communicate certain unspoken messages; Champagne or First Growths from Bordeaux for example, are able to confer on the drinker, feelings of sophistication and success. Few products capture people's imagination in this way.

The dramatic growth in the New Zealand wine industry since the 1990s reflects how, as a nation, we have adopted wine as the preferred beverage. In turn, we can see that overseas wine lovers want to enjoy this taste of New Zealand in a glass in their own home. As the plantings of vines have grown, so too has the quality of our wines increased; viticulturists and winemakers have become more skilled at their craft and the wine styles have grown more diverse.

My goal in writing this book is to offer a useful overview of the New Zealand wine regions as well as provide a textbook to be used as a resource for the NZQA wine education unit standards. The maps, winemaking and wine tasting vocabulary, background history plus Kevin Judd's evocative photographs give depth to this narrative. I am also intrigued by the recent work on wine preference profiling by Tim Hanni MW and have included his liberating message that we should feel free to like the wines we like and not feel pressured to learn to like those wines that wine commentators and educators tell us we should like!

Chapter One looks at New Zealand's wine regions with regional maps and details of their climatic influences, soil types and key grape varieties. It also outlines the role of New Zealand Winegrowers, the national organisation, with a brief look at international investment in New Zealand wine.

Chapter Two covers how grapevines are cultivated and the environment that makes them thrive. This includes vine propagation, viticultural techniques, an overview of the vineyard cycle, as well as information on sustainability, organic and biodynamic principles.

Chapter Three and **Chapter Four** discuss the key grape varieties grown in New Zealand. They cover their background, aromas, tastes, styles, viticulture and regions, as well as recommendations of wines to taste that reflect these different styles.

Chapter Five looks at how to make wine and presents processes for white, red, rosé, dessert and sparkling wines. Winemaking terms are explained as well as wine label information, closures, ageing and faults in wine.

Chapter Six discusses how to build your skills at identifying flavour. What we come to realise is that this is personal and about how a wine feels to you. This chapter looks at the formal wine-tasting process.

Chapter Seven tells the story of how the wine industry got to where it is today, with a brief historical overview of hospitality and wine, including key issues from phylloxera to liquor licensing, significant people and a timeline of historic events.

Celia Hay

As a most basic definition, wine is simply a beverage made from the fermented juice of grapes.

If you leave a bunch of grapes long enough in a bowl in your warm kitchen, you will see them slowly shrivel, develop mould from the yeasts on the skin, eventually split and start to bubble. This is the beginning of the fermentation process. Yeasts are microscopic organisms that live naturally alongside grapes in the vineyard and winery. They are carried by wind to rest on the grape's skin. The yeast consumes the sugars in the grapes and turns them into alcohol and carbon dioxide gas. All fruit will ferment, but the interesting thing about grapes is that they have the ability to be transformed into a beverage that has many personalities: it can be aged for decades or bottled so that carbon dioxide is captured within to give a natural sparkle.

Wine is usually made from one or more varieties (often called varietals) of the species *Vitis vinifera*. Wine can also be made from other species of grape or from hybrids, created by the genetic crossing of two species. There are over 3000 *Vitis vinifera* varieties in existence, which offers endless choices for the winegrower in terms of style and production methods.

Wine is a fascinating blend of nature and nurture. Different varieties of grapes and strains of yeasts produce diverse styles of wine. The result reflects a range of influences, including the growing environment, biochemical processes during fermentation and the grower's and winemaker's intervention in this process.

The fermentation process

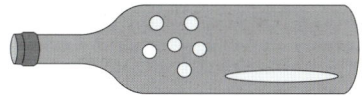

SUGAR + YEAST $\rightarrow$ ALCOHOL + CO$_2$

STYLES OF WINE

The term style can be confusing because it can mean many different things; for instance, a wine may be described as light-bodied, medium-bodied or full-bodied, which relate to how it feels on your palate. A wine may also be described as reflecting the characteristics of where it has grown: for example, *'wine made in the style of Marlborough Sauvignon Blanc'.*

In the broadest sense, wine is categorised in the following ways:

Light wines

Most wine falls into this category and this is what you would normally think of when someone mentions 'wine'. In the past, people often referred to this as 'table wine'; however, because of the European Union classification system, the phrase table wine is now associated with wine of lower quality.

- Light wines are still (not sparkling) and can be bone-dry on the palate through to sweet wines.
- Alcohol level ranges from 8% to 15% alcohol.
- Many wines are named after the region where they are produced, e.g. Chablis from the village of Chablis in France, Barossa Shiraz from South Australia, Central Otago Pinot Noir.

Sparkling wines

In sparkling wine, bubbles of carbon dioxide gas are trapped in the bottle. There are a number of ways to achieve this and certain phrases on a bottle should indicate the method.

- Made in the traditional method or *méthode traditionnelle* is when a still wine has fermented for a second time in the bottle in which the wine is sold, e.g. Champagne, Cava from Spain, and New Zealand method traditional such as Cloudy Bay Pelorus or Quartz Reef Méthode Traditionnelle.
- Tank-fermented sparkling wine is a cheaper process and this is reflected in the price that you pay, e.g. Asti and Prosecco from Italy, or Lindauer Brut.

Fortified wines

These wines have been 'fortified' or made stronger by the addition of extra alcohol during the winemaking process. They are often sweet and many wines have the ability to age for decades.

- The alcohol levels can be between 15% and 22%, e.g. Sherry from Spain and Port from Portugal.

EUROPEAN UNION FRAMEWORK

In 2009, the European Union introduced a new system of geographical indications to promote and protect names of quality agricultural products and foodstuffs including wine.

Protected Designation of Origin (PDO) In order to receive the PDO status, the entire product must be traditionally and entirely manufactured (prepared, processed and produced) within the specific region and so acquire unique properties.

- Reflects the quality and characteristics of the wine due to a particular geographic environment and production area.
- Grapes must be *Vitis vinifera* of permitted grape varieties, and 100% of the grapes must come from the area.
- Production must take place in the area.
- For wine, it is still common for PDOs to use the traditional names that were designated under the previous framework. For instance in France, labels still include the Appellation Contrôlée designation for that particular wine.

Protected Geographic Indication (PGI) In order to receive the PGI status, the entire product must be traditionally and at least partially manufactured (prepared, processed or produced) within the specific region and so acquire unique properties.

- Reflects the specific quality and reputation of an area.
- Grapes must be *Vitis vinifera* or a hybrid, and 85% of grapes must come exclusively from the area.
- Production must take place in the area.

All other wines: This category replaces 'table wine'. These are wines without a specific geographic indication and can be a blend of wine from several EU countries. They may show a vintage and can be varietally labelled.

Upper Brancott Valley with Blairich Mountains.

Terroir

This word of French origin is used to describe the set of environmental influences that affect the growing of certain products, including grapes. The expression of these influences in the product helps tie them to a particular location by attributing them with certain qualities.

In France, the concept of terroir has evolved over generations to be used to describe a sense of place, embracing the history of a growing region including its traditional practices, the influence of the soil structure, local climate (sun, rain) and weather patterns that may enable a product to thrive. In wine, this translates to the following.

- Traditional grape varieties indigenous to a specific region.
- The profile of the soil, e.g. clay, limestone, schist etc.; water availability and how the soil 'feeds' the vines with nutrients.
- How the vines are grown, trained and tended.
- Aspects of the growing site in relation to the sun.
- Traditional winemaking practices, e.g. use of indigenous yeasts, fermenting in oak barrels, extended barrel ageing for a specific period, and so on.

In France, the Appellation d'Origine Contrôlée (AOC) system was established in 1935 to classify areas according to their ability to produce quality wine and to reflect and codify the attributes of the terroir of a particular area. This model has been extended to all wine-producing countries of the European Union, although it is called by different names in these counties. In 2009, a new system of Geographic Indication was introduced for the European Union. (See box on p. 4)

NEW ZEALAND TERROIR

As New Zealand is a relatively new wine-producing country, our winegrowers are constantly debating the concept of terroir and its relevance to their own wines. As the vines grow in age and the sub-regions become recognised clusters of quality, the terroirs of New Zealand will become more identifiable to the consumer.

Terroir, in the first instance, is about the land and soil, its drainage, the aspect, topography, the climate in which grapes grow – rainfall, sunlight, winds and temperatures, among other factors. Many people describe terroir as 'the soul of the land'.

The vineyard practices, including irrigation of vines and winemaking process influence this. It is claimed, however, that when wine is made to a recipe, managed and corrected by the addition of chemicals and additives, the expression of the terroir, the individual character of a specific vintage, is lost. The accuracy or otherwise of this claim forms part of an ongoing debate.

Springtime, Hans Herzog Estate, Marlborough.

Production Models:
Three wine stories

Not all wines are created equal. I like to illustrate this by telling the tale of three wine production models.

Story one consists of a select group of wineries that have developed and fostered an international reputation for their wines and drive the fine wine, luxury end of the wine industry.

These wines are expensive, highly sought after and often viewed more as an investment that will appreciate in value than a beverage to enjoy with dinner. Most of these wines are from Bordeaux and Burgundy with a handful from Italy, Spain, California, Chile and Australia.

Story two consists of wine production that is increasingly part of the international processed food industry and sold through supermarket chains.

These wine companies have embraced science and technology to produce wine that appeals to a broad customer base and sell profitably at the price point that consumers wish to pay. Increasingly, these wines are well-known international brands. Their labels are engaging and easily communicate their attributes through carefully chosen words and imagery, including wine medals and awards.

Story three consists of the small to medium producer who has inherited or established a winery because of an enormous passion to make wine or through family tradition.

This is the artisan producer who only makes small quantities of wine, using science as well as instinct to fashion a distinctive style of wine. These wines, because of their hand-made nature, are more expensive and as a consequence fail to get selected by supermarkets. These wines struggle to attain brand recognition because they lack the marketing budget and scale of production to achieve economies of scale.

As a consumer, it is good to learn how to identify these different models. In most instances, price is the key as low prices tend to equate to large-scale production of lower quality wine. In general, when you buy a discounted wine at the supermarket, it's nice to feel that you have got a bargain, but remember that price reflects quality.

And, as you learn more about wine, the bad news is that you will start spending more on wine. This is because you understand and appreciate the work that goes into making quality wine.

Millton's biodynamic barrels.

NEW ZEALAND WINE REGIONS & KEY VARIETIES

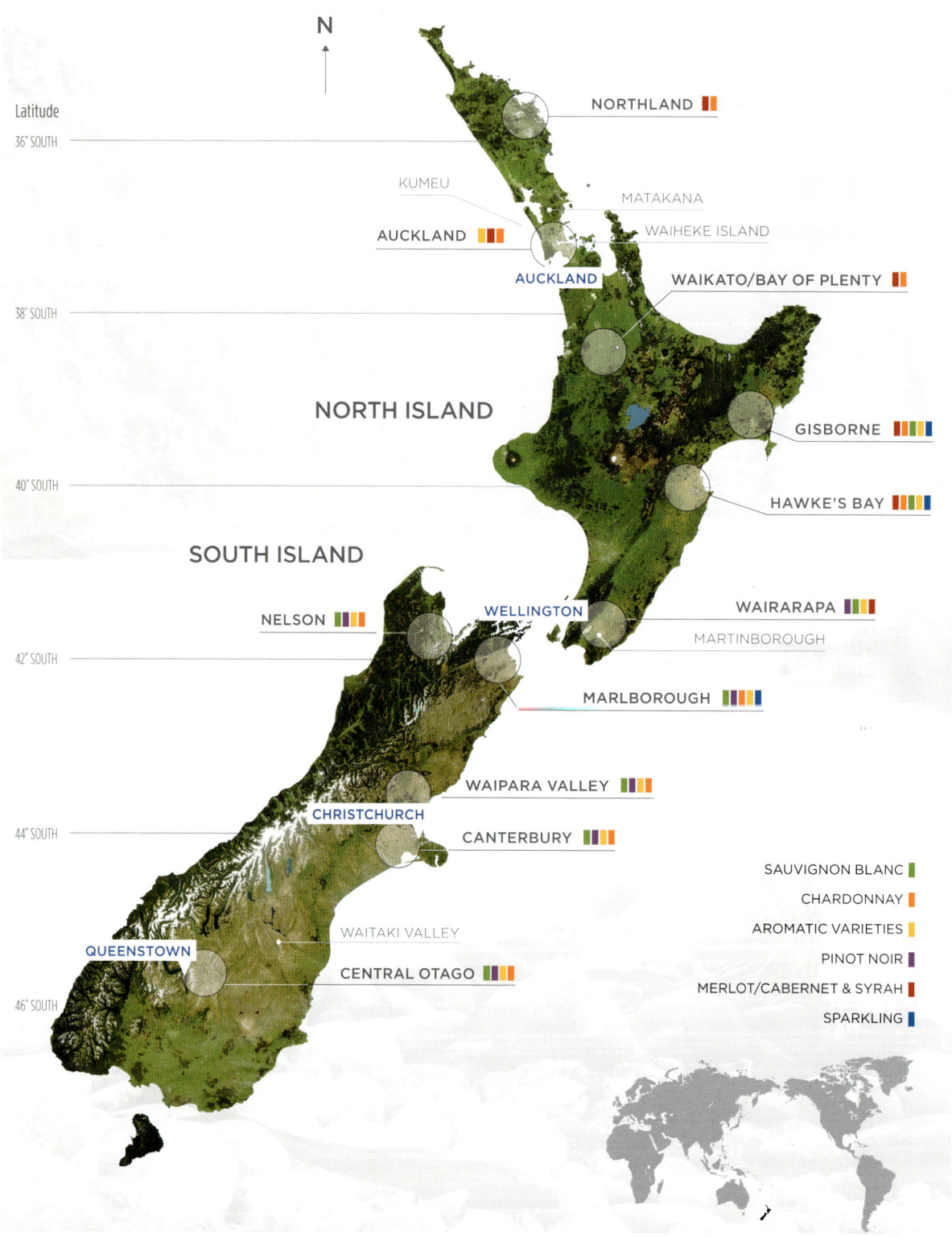

N

Latitude
36° SOUTH

NORTHLAND

KUMEU

MATAKANA

AUCKLAND

WAIHEKE ISLAND

AUCKLAND

WAIKATO/BAY OF PLENTY

38° SOUTH

NORTH ISLAND

GISBORNE

40° SOUTH

HAWKE'S BAY

SOUTH ISLAND

NELSON

WELLINGTON

WAIRARAPA

MARTINBOROUGH

42° SOUTH

MARLBOROUGH

WAIPARA VALLEY

CHRISTCHURCH

44° SOUTH

CANTERBURY

WAITAKI VALLEY

QUEENSTOWN

CENTRAL OTAGO

46° SOUTH

SAUVIGNON BLANC

CHARDONNAY

AROMATIC VARIETIES

PINOT NOIR

MERLOT/CABERNET & SYRAH

SPARKLING

Source: New Zealand Winegrowers (for more information see www.nzwine.com).

CHAPTER ONE:

New Zealand wine regions

New Zealand's long, thin islands lie approximately between the latitudes of 36 to 46 degrees south of the equator. New Zealand is considered to have a maritime climate due to the close proximity of the sea and its moderating influence. This means that the seasons are less extreme. New Zealand is a windy country and generally there is always cloud with regular rainfall. Central Otago, which is furthest from the sea, is considered to have a continental climate.

Auckland, Gisborne, Hawke's Bay and more northern wine regions have a moderate climate and are therefore more suited to varieties such as Syrah, Cabernet Sauvignon, Cabernet Franc and Merlot, which require warmer temperatures to ripen.

The Wairarapa and all of the South Island are considered cool-climate regions. They are often described as having a *'brightness of fruit'.* Grape varieties that excel in cool climates include Sauvignon Blanc, Chardonnay, Riesling and Pinot Noir. These regions tend to have a longer hang time for the grapes that results in the development of more intense fruit flavours and retention of higher levels of acid. Cool-climate wines are very fashionable.

New Zealand and Australia also have a very high light intensity with much higher levels of ultraviolet light than the rest of the world. Experts believe that this may have a major influence on why our wines are so 'varietally vibrant'[1] with an intense, upfront, fruit-forward character.

Another characteristic of New Zealand's leading wine areas, similar to the great wine regions of the world, is the common attribute of river terraces, carved out from ancient river systems that consist of low fertility soils made up of stones, gravel and sand which drain easily and do not hold onto moisture.

NIWA maps

Viticulturists consider many elements when selecting vineyard sites including classifications such as degree days, which measure the heating and cooling, of a particular site or area.

The National Institute of Water and Atmospheric Research (NIWA) is a Crown Research Institute and these maps help to give a visual overview of important aspects of New Zealand's climate.

For more information visit http://www.niwa.co.nz

MEDIAN ANNUAL SUNSHINE HOURS – PAGE 11

Winegrowers often talk of the sunshine hours of a particular area. The measurement of sunshine is also an indicator of cloudiness in an area.

This map clearly shows that Gisborne and Marlborough have higher levels of sunshine than Auckland and Northland even though the latter two regions lie closer to the equator and have higher daily temperatures.

MEDIAN ANNUAL RAINFALL – PAGE 12

This map shows the average rainfall each year. It is useful to see how localised rain can be.

Central Otago clearly has the lowest rainfall, with pockets of Marlborough and Canterbury also receiving less than 500 millimetres per year. The West Coast, by contrast, gets over 4000 millimetres or 4 metres of rain.

MEDIAN ANNUAL TEMPERATURE – PAGE 13

New Zealand has a moderate, temperate climate. The hottest regions are those at the top of the North Island and coastal Gisborne and Hawke's Bay, with an average temperature of 14–16°C. By contrast, Central Otago averages 10–12°C.

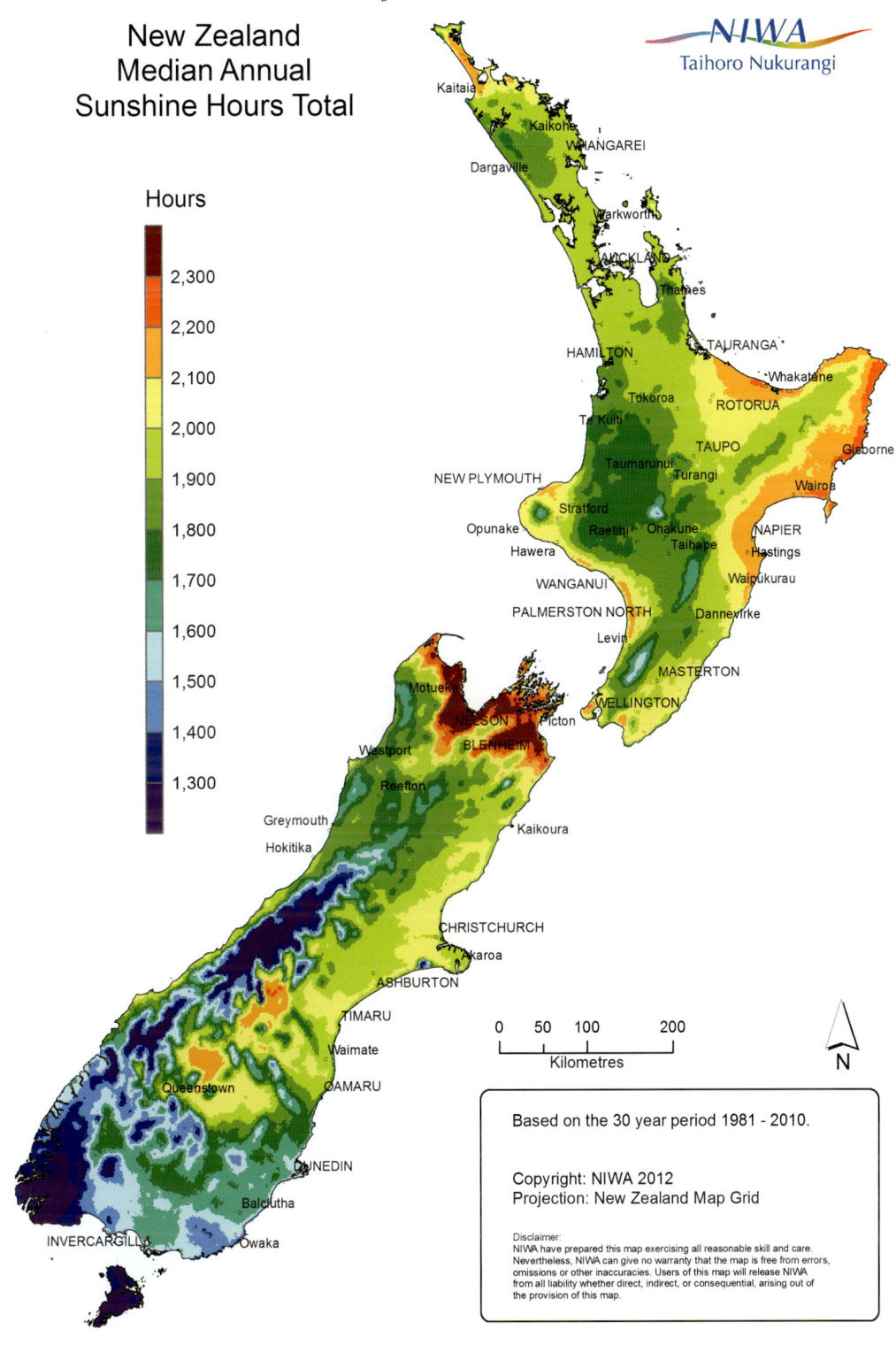

New Zealand
Median Annual
Sunshine Hours Total

Hours

2,300
2,200
2,100
2,000
1,900
1,800
1,700
1,600
1,500
1,400
1,300

Based on the 30 year period 1981 - 2010.

Copyright: NIWA 2012
Projection: New Zealand Map Grid

Disclaimer:
NIWA have prepared this map exercising all reasonable skill and care.
Nevertheless, NIWA can give no warranty that the map is free from errors,
omissions or other inaccuracies. Users of this map will release NIWA
from all liability whether direct, indirect, or consequential, arising out of
the provision of this map.

0   50   100   200
Kilometres

N

MEDIAN ANNUAL RAINFALL

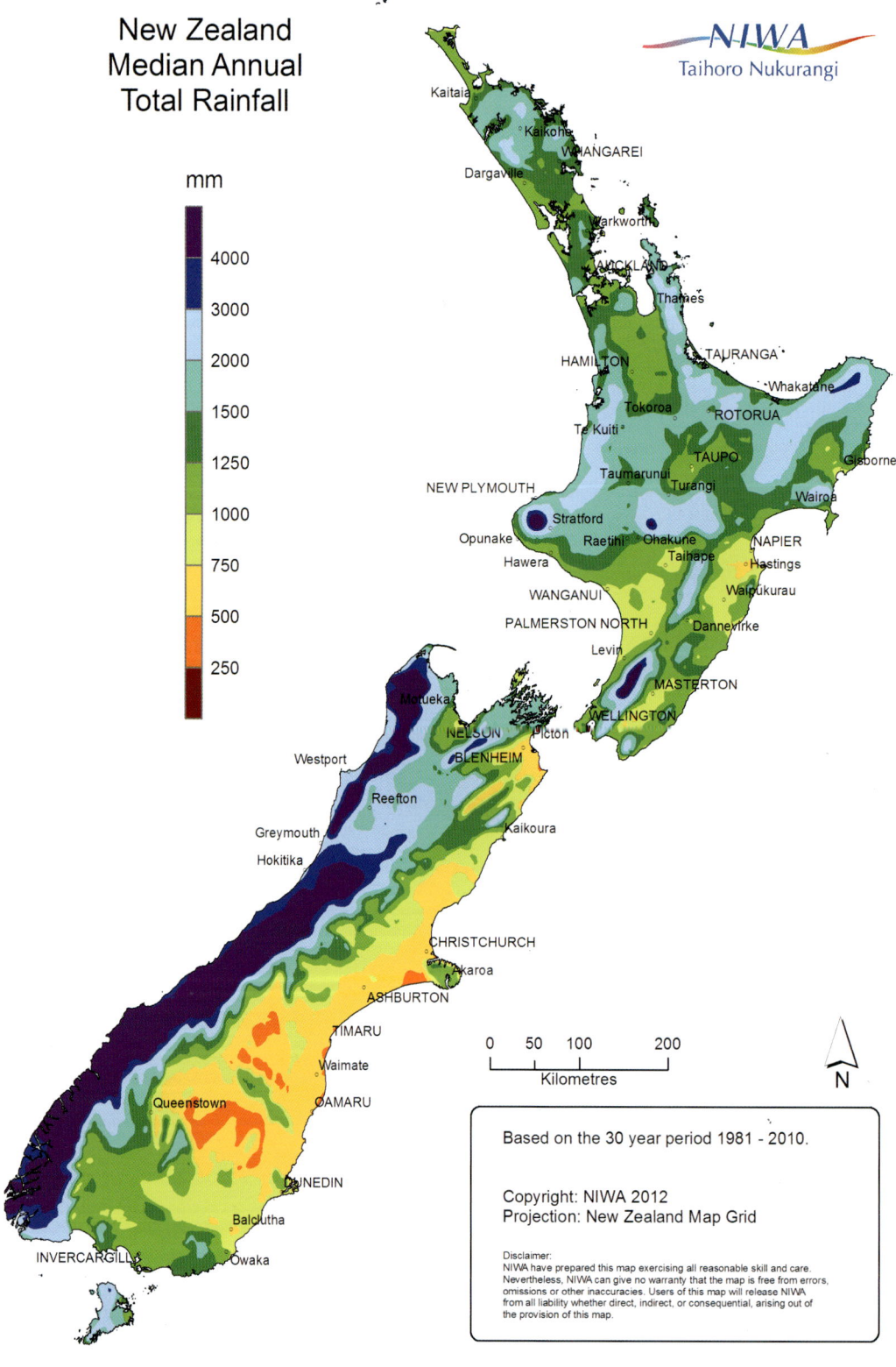

New Zealand
Median Annual
Total Rainfall

mm

4000
3000
2000
1500
1250
1000
750
500
250

Based on the 30 year period 1981 - 2010.

Copyright: NIWA 2012
Projection: New Zealand Map Grid

Disclaimer:
NIWA have prepared this map exercising all reasonable skill and care.
Nevertheless, NIWA can give no warranty that the map is free from errors,
omissions or other inaccuracies. Users of this map will release NIWA
from all liability whether direct, indirect, or consequential, arising out of
the provision of this map.

MEDIAN ANNUAL TEMPERATURE

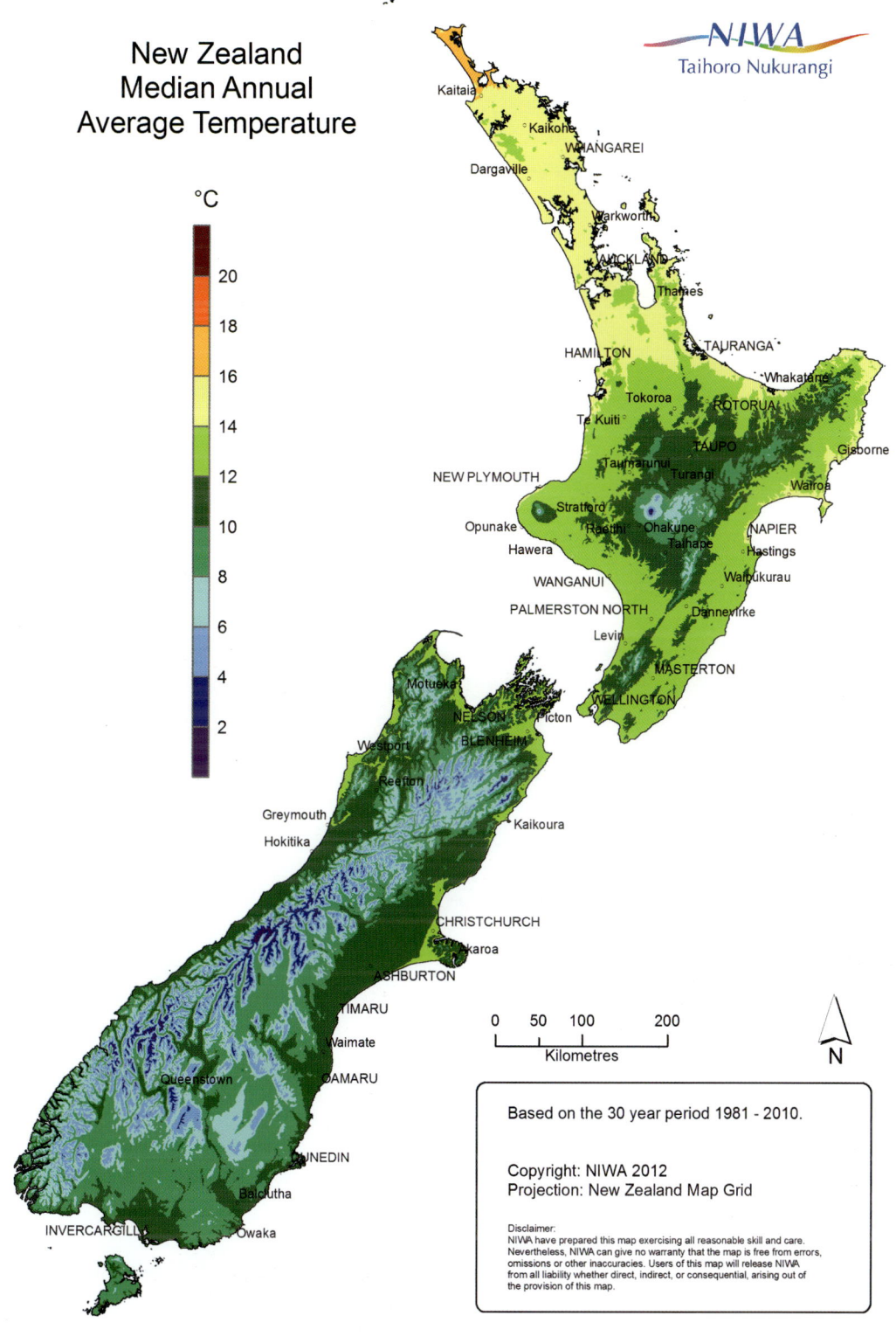

New Zealand
Median Annual
Average Temperature

°C

20
18
16
14
12
10
8
6
4
2

Based on the 30 year period 1981 - 2010.

Copyright: NIWA 2012
Projection: New Zealand Map Grid

0   50   100   200
Kilometres

N

Auckland

Man O' War Lone Kauri Vineyard, Waiheke Island.

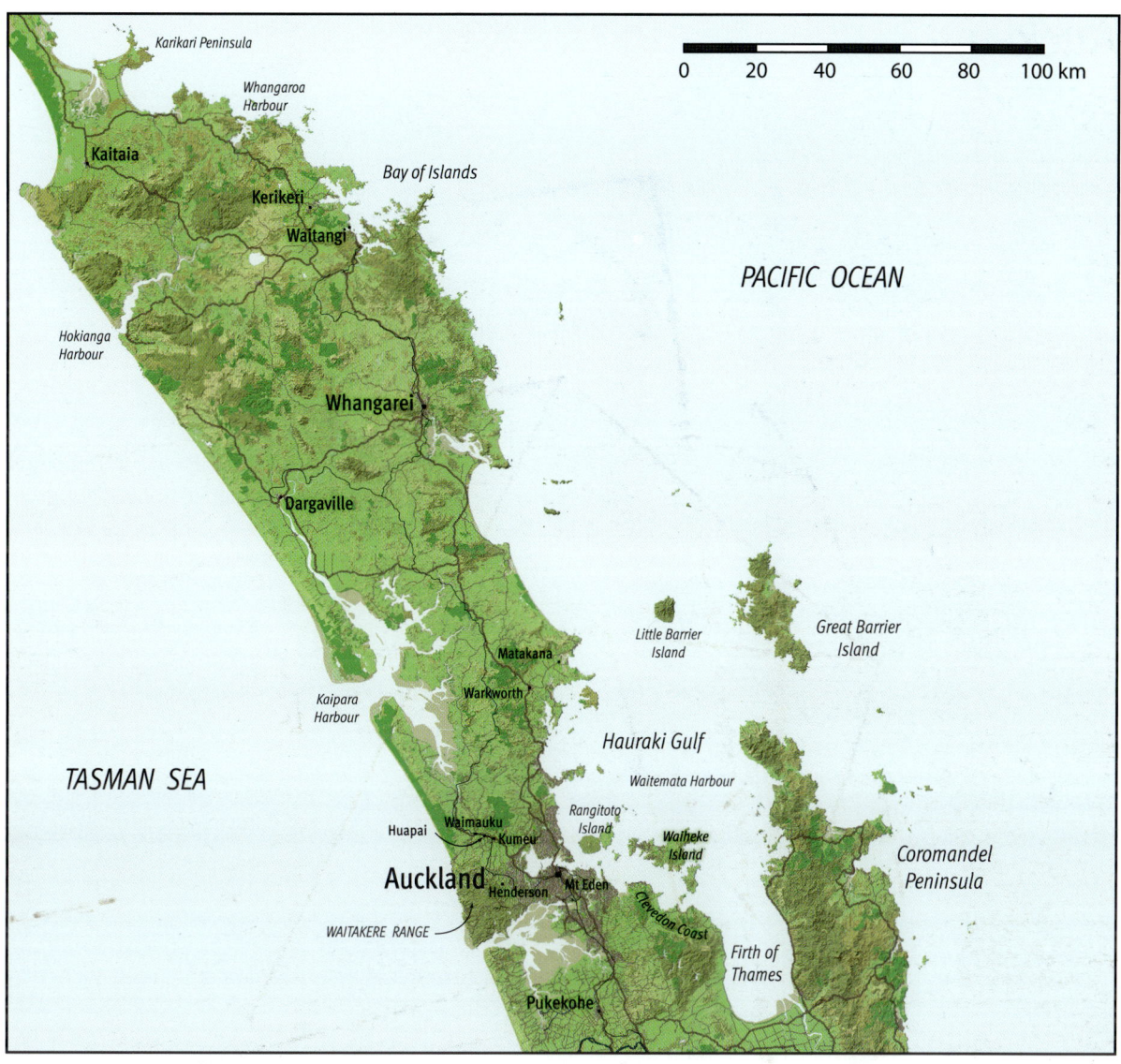

Auckland is a diverse wine region that extends way beyond the boundaries of the City of Auckland, New Zealand's largest city. While the area of vine plantings has been overtaken by other regions, historically the Auckland wineries led the charge for innovation in the sector and growth of vineyards into Marlborough during the 1970s. Although Auckland's contribution to the national vineyard area is small, it remains the headquarters for many of the largest wineries and also many small, boutique producers.

HISTORY

West Auckland is the home of many Dalmatian/ Croatian families who established small vineyards, often alongside market gardens, in the early twentieth century. Babich was established in 1916. Selaks and I Yukich and Sons were established in 1934 and went on to become Montana Wines in 1961. Others included Nobilo's (1943), Delegat's (1947) and Villa Maria (1961). The exception to the Dalmatian dominance was Corbans Wines which was established in 1902 by Assid Corban, originally from Lebanon. By 1960, there were nearly 90 wineries in the Henderson-Kumeu area.[2] Many of these pioneering families have gone on to be the innovators and leaders of New Zealand's modern wine industry.

In the greater Auckland area, Bordeaux blends have come to dominate as the warm climate ensures that, in most years, the grapes can ripen and produce full-bodied red wines. Recent plantings of Syrah are producing outstanding wines, especially on Waiheke Island. Top Chardonnays are also produced.

CLIMATE

The climate of the greater Auckland area is maritime and dominated by the Pacific Ocean to the east and the Tasman Sea to the west. In New Zealand's cool climate, Auckland is considered to be at the warmer end of this scale. Never far from either coast, the weather is moderately warm with good cloud cover, relatively high rainfall and humidity.

SUB-REGIONS

West Auckland: Henderson, Kumeu, Huapal, Walmauku

The sub-regions of Henderson, Kumeu, Huapai and Waimauku are located in West Auckland, approximately a 30-minute drive from downtown Auckland. The high humidity and rainfall creates many viticultural challenges. Despite this, Kumeu River Wines, under the direction of Michael Brajkovich MW, have forged an international reputation for their range of single-vineyard and estate Chardonnays.

Climate: Moderately warm, some frosts with relatively high rainfall and humidity.
Soil: Heavy clay soils.

Predominant grapes: Chardonnay, Pinot Gris, Merlot, Syrah.

Try these wines: Cooper's Creek Huapai Montepulciano, Soljans, West Brook Waimauku Pinot Gris.

Waiheke Island

Situated in the Hauraki Gulf, Waiheke Island, the trip takes 40 minutes on regular commuter ferries that depart hourly from downtown Auckland. This voyage, along with a tour to three or four vineyards, has become one of the must-do experiences for visitors to Auckland.

Like West Auckland, Waiheke Island has played a significant role in the establishment of New Zealand's modern wine industry. Small plots on rolling coastal land enabled important experimentation.

Serious planting on Waiheke started in 1978 when Jeanette and Kim Goldwater planted one hectare of Cabernet Sauvignon. They were enthusiasts but had no equipment, no experience and no idea of how to make wine.[3] Their first commercial vintage of 300 cases was released in 1983. In the years that followed, they planted more vines of different varieties and flourished. In 2011, the Goldwaters gifted the vineyard to Auckland University for use as part of its Wine Science programme. Stephen White planted Stonyridge in 1982 and, within a decade, Stonyridge Larose became New Zealand's most celebrated and expensive Bordeaux blend.

Climate: Moderately warm and windy with salt spray coming off the sea, rainy.

Soil: Widespread undulating hills with clay soils.

Predominant grapes: Cabernet Sauvignon, Merlot, Chardonnay, Syrah.

Try these wines: Thomas and Sons Cabernet Sauvignon, Man O' War Valhalla Chardonnay, Mudbrick Merlot Cabernet, Obsidian 'The Mayor' (Cabernet Franc, Petit Verdot, Malbec), Te Motu Kokoro, Te Whau Chardonnay.

Matakana

Matakana is one hour's drive north of Auckland. Most vineyards are located within a few kilometres of the sea. Matakana is a popular weekend destination from Auckland and the wineries have a loyal local following.

Climate: Moderately warm with relatively high humidity and rainfall, which suits red varieties.

Soil: Heavy clay soils which can hold moisture.

Predominant grapes: Merlot, Cabernet Sauvignon, Syrah, Chardonnay.

Try these wines: Providence, Brick Bay.

Clevedon

Clevedon is located 40 minutes' drive south of Auckland and is closely associated with pastoral farming and horse breeding.

Climate: Moderately warm with relatively high humidity and rainfall, which suits red varieties.

Soil: Heavy clay soils that can hold moisture.

Predominant grapes: Merlot, Cabernet Sauvignon, Syrah, Chardonnay.

Try these wines: Puriri Hills Reserve (Merlot, Carménère).

KEY WINE STYLES IN AUCKLAND

Chardonnay

Kumeu River Wines is considered to make the most highly regarded wine in the greater Auckland region. They produce five different styles of Chardonnay with their premium wines from single-vineyard sites. Their goal is to make wines in the Burgundy style using indigenous yeast, lees-ageing and malolactic fermentation. Villa Maria, from its headquarters in Manukau, also produces outstanding Chardonnay and Gewurztraminer from the volcanic soils of the Ihumatao vineyard.

Bordeaux blends: Merlot, Cabernet Sauvignon and Cabernet Franc

Some years these grapes struggle to ripen when the growing season is cool and wet. Premium wines are consistently produced, such as Stonyridge Larose, Destiny Bay Magna Praemia, Providence.

Syrah

This famous French grape from the Rhône Valley has grown in popularity, particularly on Waiheke Island, where it makes intense fuchsia-coloured wines with an aroma of ripe berries and pepper. Premium wines include The Hay Paddock, Man O' War, and Kennedy Point, Expatrius, Passage Rock.

Pinot Gris

Pinot Gris is the second most planted white wine in the Auckland area. It produces wines with citrus, stonefruit and pear aromas. Excellent examples include: Cable Bay, Batch Winery, Man O' War.

Sauvignon Blanc

Waiheke Island also offers some great examples of Sauvignon Blanc: Expatrius, Man O' War.

Brick Bay Vineyard, Matakana.

Northland

Wine production in Northland is small, considering that the Bay of Islands is where grapes were first planted in New Zealand. In 1819, the missionary Reverend Samuel Marsden planted vines at the small Church of England mission of Rangihoua.

Climate: Relatively hot and humid and with frequent falls of rain; closeness of the sea has an important tempering influence on the climate.

Soil: Mainly clay-rich loam over compact clay.

Predominant grapes: Cabernet Sauvignon, Merlot, Pinotage, Chardonnay, Pinot Gris.

Try these wines: Longview Estate, Marsden Estate.

Waikato and Bay of Plenty

WAIKATO

History

The Waikato region is primarily known as a centre of the diary industry; however, in 1886 a government research station was established at Te Kauwhata. This was initially to research agricultural crops, but in 1901 Romeo Bragato set up the viticultural division. Wine was produced here in small quantities from 1903.

In the 1980s, the Te Kauwhata winery was branded as Rongopai Wines and was bought by Babich in 2007.

Climate: Warm with high humidity and high annual rainfall.

Soil: Clay loam soils.

Predominant grapes: Chardonnay, Cabernet Sauvignon.

BAY OF PLENTY

Two of New Zealand's best-known wineries, Mills Reef and Morton Estate, are based in this region although both own significant vineyards in Hawke's Bay and Marlborough.

Wine production is small and focused mainly on Chardonnay, Cabernet Sauvignon and Sauvignon Blanc. The region enjoys a moderately warm climate and fertile soils by New Zealand standards, while still having a coastal influence.

Climate: Moderately warm with coastal winds.

Soil: Fertile clay.

Predominant grapes: Chardonnay, Cabernet Sauvignon.

Gisborne

Millton Naboth's Vineyard, looking north across Poverty Bay Plains.

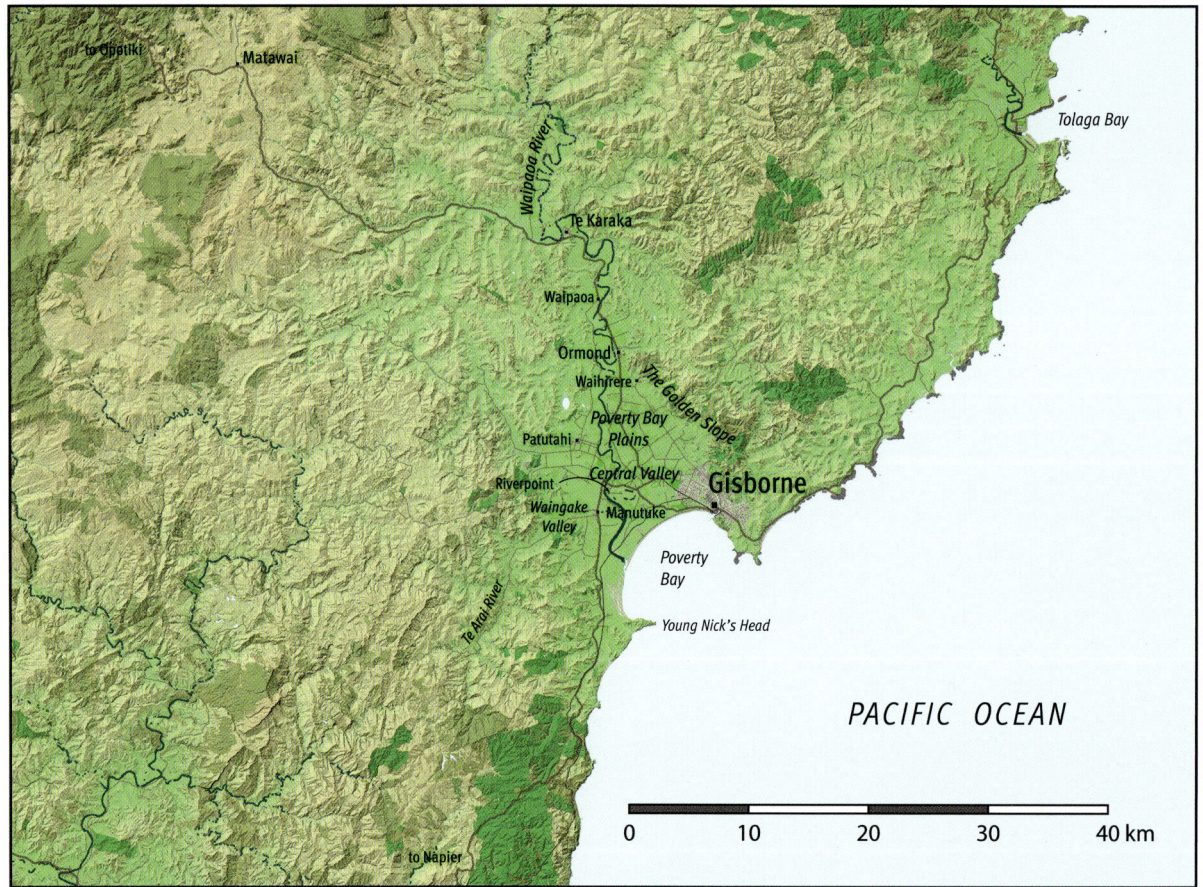

Gisborne is known for its high sunshine hours and warm climate, although this is tempered by bursts of rain. Increasingly, Gisborne is considered a white-wine region where aromatics thrive. It is one of the first regions to start the harvest each year, often two weeks earlier than Hawke's Bay or Marlborough.

HISTORY

Gisborne is the first city in the world to see the sun each day and is New Zealand's fourth largest wine region. Small pockets of grapes were grown from the nineteenth century, but the region really took off when Corbans began to plant Muller-Thurgau in the 1960s and was quickly followed by other large wine companies. During the 1970s Gisborne was the largest wine-growing region in New Zealand, with Muller-Thurgau the most planted vine. Vines flourished until the mid 1980s when growers were given the opportunity to receive compensation and pull out vines as part of the government's vine-pull scheme, which followed a wine glut. Many replaced their Muller-Thurgau with Chardonnay.

Today the region is associated with the production of Lindauer and bulk red and white wines, such as Wohnsiedler, named after an early winemaker from the 1910s. A number of artisan producers have emerged and gone on to build national reputations for their wines.

Chardonnay remains the most important grape, but aromatic grapes such as Pinot Gris, Gewurztraminer and Viognier are also producing highly regarded wines.

SUB-REGIONS

Patutahi

Famous for its top Gewurztraminer, Patutahi is an inland area north-west of Gisborne city and considered to be warmer and drier than the other sub-regions.

Climate: Lower rainfall than other nearby regions and cooler.

Soil: Well-draining clay, silt.

Predominant grapes: Gewurztraminer, Chardonnay.

Try these wines: Matawhero Merlot.

Manutuke

Manutuke was originally planted in the 1890s and is located south of Gisborne city and close to the coast.

Climate: Warm, with sea breezes; ideal conditions for the formation of *Botrytis cinerea*.

Soil: Well-drained sandy, silt soils with some heavier clays.

Predominant grapes: Chardonnay.

Try these wines: Millton Clos de Ste Anne.

Ormond

Ormond is north of Gisborne city and where the first commercial vineyards were established. Today it has some of the best vineyard sites producing single-vineyard wines. It is also where the 'Golden Slope' is located. This is a 10-kilometre elevated strip, facing south-west with 20–30 centimetres of light black topsoil. The Golden Slope has produced many top Chardonnays.

Climate: Warmer, slightly drier with silt-loams prevailing.

Soil: Gently sloping, free-draining, sandy escarpment with limestone-influenced topsoil.

Predominant grapes: Chardonnay, Gewurztraminer.

Try these wines: Matawhero Chardonnay, Huntaway Gisborne Viognier, TW Estate Unoaked Chardonnay.

KEY WINE STYLES

Full-bodied Chardonnay and aromatic white wines made from Gewurztraminer, Pinot Gris, Muscat and Viognier are the leading styles. Gisborne is still a source of grapes for wineries located elsewhere in the country.

Chardonnay and Chenin Blanc

The Millton Vineyard, owned by James and Annie Millton, was the first vineyard in New Zealand to gain organic certification. It now has biodynamic certification as well. James Millton has been the leader in this movement among New Zealand wineries. Millton is best known for its Chenin Blanc, made in a number of styles and premium Chardonnay from the Clos de Ste Anne vineyard located on a steep north-facing slope.

Gewurztraminer

Nick Nobilo, former head winemaker and CEO of Nobilo Wines, owns Vinoptima, a boutique vineyard specialising in Gewurztraminer. He firmly believes that Gewurztraminer is the unsung hero of New Zealand wine, showing complex and exotic aromas and suited to a number of winemaking styles. The premium quality and purity of his wines are an exciting development for the Gisborne region.

Hawke's Bay

Te Mata Woodthorpe Vineyard, Dartmoor Valley, Hawke's Bay.

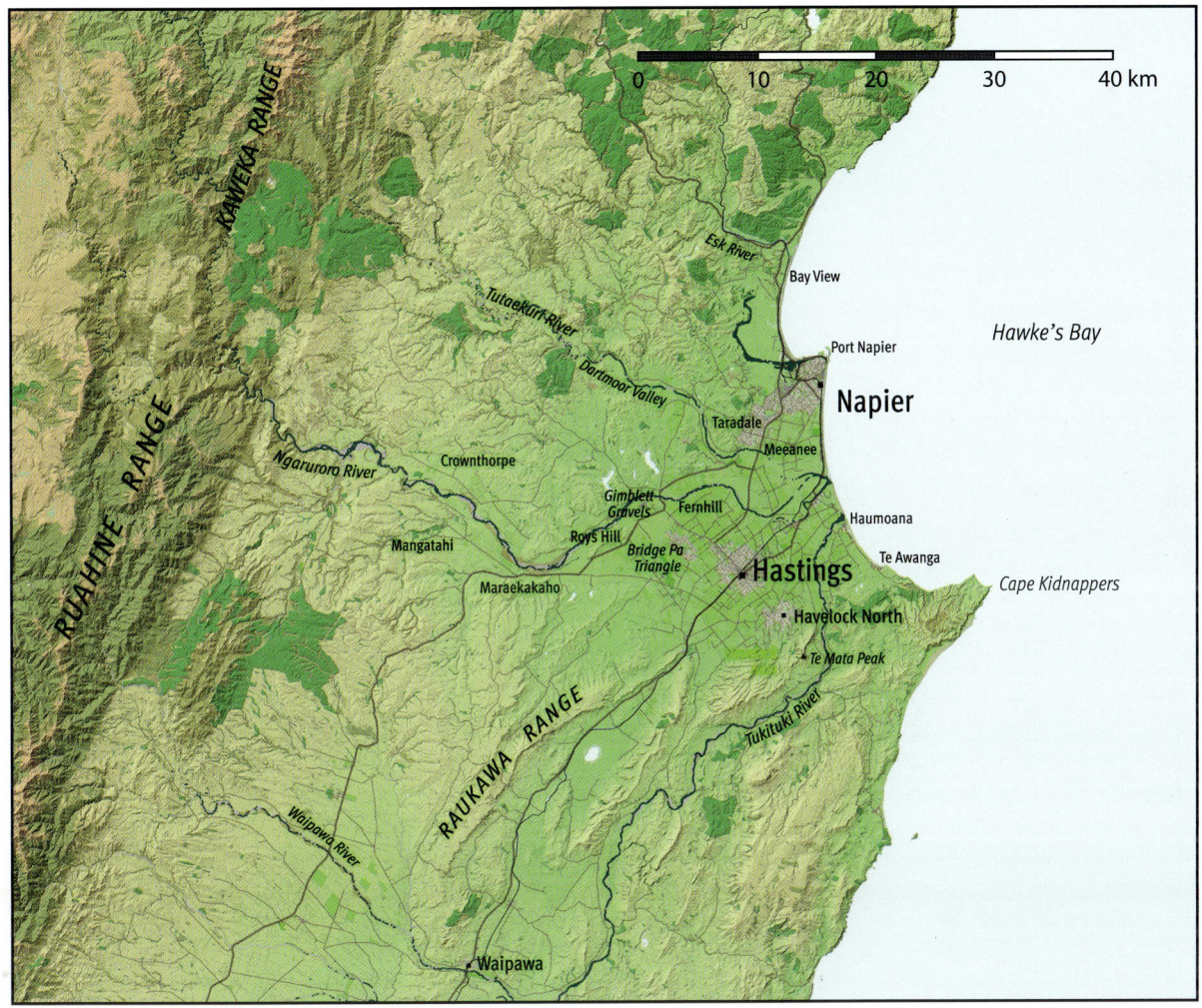

Hawke's Bay is New Zealand's second largest wine region with a rich history of grape growing and horticulture, going back to the 1850s. The long sunshine hours make Hawke's Bay well-suited to the later ripening grape varieties such as Cabernet Sauvignon, Merlot and Syrah. Hawke's Bay also produces first-rate Chardonnay in a variety of styles. Syrah is growing in influence and while plantings remain small, the quality of Syrah produced seems to increase as every vintage passes.

From its early days, Hawke's Bay has built its reputation as a red wine region. Cabernet Sauvignon, Merlot, Cabernet Franc and Malbec, blended together in different ways and referred to as Bordeaux blends, gained a loyal following. By the 1990s, Merlot had started to replace Cabernet Sauvignon as the dominant partner of the blend.

Vines are found throughout Hawke's Bay and planted on a wide range of soils and in diverse conditions. The main grape growing areas are influenced by the four major rivers that dominate the region with their valleys and river terraces.

Historically, there was no pattern about where vines should be grown, however plantings since the 1990s have focused on sites with lower fertility.

HISTORY

In 1851, a mission station was established by Marist brothers from France in Meeanee, now a suburb of Hastings. They planted vines for sacramental purposes but by 1870 were selling small quantities of wine. About 1890, Henry Tiffen of Greenmeadows Station followed by Bernard Chambers of Te Mata Station planted vines. At the time of Tiffen's death in 1896, there were approximately ten hectares planted. In 1897, the Marists went on to purchase over 300 hectares from Tiffen's estate and eventually moved the mission to this land which remains its site today. Anthony Vidal, a Spanish immigrant, became the first commercial winemaker in the region in 1905. Vidal, under the stewardship of Villa Maria, continues to be an important Hawke's Bay winery. Tom McDonald crafted his first Cabernet Sauvignon wine in 1949 at what is now known as the Church Road winery. In the 1960s McWilliams dominated Hawke's Bay production with a series of commercially successful wines. The vineyard area more than doubled in the 1970s with plantings of Muller-Thurgau. By the 1990s the vineyard area had doubled again.

CLIMATE

The Ruahine Range to the south and the Kaweka Range to the north, line the western boundary of Hawke's Bay. These mountains provide a rain shadow protecting the region from westerly rain and ensuring generally, a long hot summer and dry autumn. To the east, the Pacific Ocean brings coastal breezes, wind and rain and moderates the climate of those areas closer to the sea. Due to the diverse range of sub-regions and their different climates, harvest dates can vary between regions by at least a month for some grape varieties.

SUB-REGIONS

The Ngaruroro River now flows north of Hastings but historically changed its current course around the Heretaunga Plains in the 1860s and again in 1931, following the Napier Earthquake. At this time the river left behind 800 hectares of shingle, stones and sand which is now known as the Gimblett Gravels.

Esk Valley

Esk Valley is the most northern sub-region and lies adjacent to the Esk River and looks east towards the sea. The best sites are on river terraces.

Climate: Moderately hot, mild frost-free winters with cooling sea breezes.

Soil: Deep silt deposited by the Esk River; some heavy soils with strong fertility.

Predominant grapes: Chardonnay, Chenin Blanc.

Try these wines: Esk Valley The Terraces (Malbec, Merlot, Cabernet Franc) proves that Bordeaux blends can thrive in Esk Valley. Esk Valley Pinot Gris.

Dartmoor Valley

The Tutaekuri River flows along the Dartmoor Valley and has large plantings on both banks including Te Mata's Woodthorpe vineyard.

Climate: Moderately hot.

Soil: Fertile clays, silt, sand.

Predominant grapes: Merlot, Chardonnay, Sauvignon Blanc.

Try these wines: Sacred Hill Rifleman's Chardonnay, Halo Chardonnay, Te Mata Cape Crest Sauvignon Blanc.

Gimblett Gravels

The Gimblett Gravels is trademarked as a wine growing district (see P.26) and is renowned for its stony soils and hot growing climate. Gimblett Gravels achieves summer temperatures often 2-3°C warmer than other areas of Hawke's Bay. With low yields, Gimblett Gravels produces wines that are full-bodied and rich in taste.

Climate: Considered the hottest and driest sub-region.

Soil: Alluvial stony gravels, silts and sands of the Ngaruroro River trap heat during the day. The gravel soils, known as the Omahu Gravels, are dry and free-draining, retaining little moisture.

Predominant grapes: Merlot, Syrah, Cabernet Sauvignon.

The Koropiko Vineyard on the Heretaunga Plains, Hawke's Bay.

Gimblett Gravels wineries include: Trinity Hill, Alpha Domus, Te Awa, Stonecroft, Unison, Craggy Range.

Bridge Pa Triangle

This is located just south of the Gimblett Gravels and considered slightly cooler.

Climate: Hot and dry in summer.

Soil: Fine sandy loams over deep coloured gravel referred to as red metal.

Predominant grapes: Merlot, Syrah, Chardonnay.

Try these wines: Te Mata Bullnose Syrah, Ngatarawa Alwyn Chardonnay, Sileni Triangle Merlot.

Havelock Hills

The hills around Havelock North include the famed Te Mata Peak which towers over the area. To the south-east of Havelock North is the Tukituki River.

In 1996 the Havelock Hills area was recognised for the significance of its natural and viticultural heritage and was designated by the Hastings District Council as the Te Mata Special Character Zone.

Climate: Moderately hot on north-facing slopes.

Soil: Sandy loam soils.

Predominant grapes: Merlot, Cabernet Sauvignon, Chardonnay.

Try these wines: Te Mata Coleraine, Te Mata Awatea, Blackbarn Chardonnay, Askerne Sémillon.

Te Awanga

Located on the coast, Te Awanga and Huamoana, further north, have built a reputation for fine white wines especially Chardonnay, Viognier and Pinot Gris.

Climate: Cooled by afternoon sea breezes.
Soil: Well-drained, shingle, sand and clay.
Predominant grapes: Chardonnay, Pinot Gris, Sauvignon Blanc.
Try these wines: Clearview Chardonnay, Elephant Hill Viognier.

Other regions

Taradale is on the outskirts of Napier and is home to the Mission and Church Road Winery. The inland areas such as the **Mangatahi Terraces** further west up the Ngaruroro River and **Central Hawke's Bay** are more elevated and therefore cooler regions. Mangatahi is considered a white wine region with Chardonnay and Sauvignon Blanc dominating. Central Hawke's Bay is further south and cooler still, with plantings of Pinot Noir and Sauvignon Blanc.

Try these wines: Lime Rock Pinot Noir from Central Hawke's Bay, Alluviale Blanc (Sauvignon Blanc and 2% Sémillon) from Mangatahi.

KEY WINE STYLES

The main grape varieties planted in Hawke's Bay are Merlot, Chardonnay and Sauvignon Blanc at around 1000 hectares each. Pinot Gris comes in next with Syrah and Pinot Noir at just over 300 hectares.

Bordeaux blends dominated by Merlot; Syrah from the Gimblett Gravels and on the hillside close by and Chardonnay are the most highly regarded wines from Hawke's Bay.

The Bordeaux blends and Syrah are generally considered full-bodied wines made from dark, ripe fruit with intense aromas and flavour. Hawke's Bay

Chardonnay is produced in a variety of styles from highly concentrated, ripe fruit aged in oak barrels to Chardonnay that is lean and mineral with restrained use of oak.

Pinot Gris is made in an off-dry as well as in a dry style. Sauvignon Blanc, made from riper fruit than is found in Marlborough, can be medium to full-bodied with tropical and melon-fruit flavours.

GIMBLETT GRAVELS

Gimblett Gravels is the trademark of the Gimblett Gravels Winegrowers Association. The area is 800 hectares consisting of gravels from the old Ngaruroro riverbed. In addition, the association requires members to own their own vineyard land and that 95% of the vineyard displays soil types associated with the river.

'To the best of our knowledge this is the first viticultural appellation in the New World where its ultimate boundary is defined by a distinct soil type boundary, no compromises, no politics'.

Red grapes account for 90% of the plantings: 35% Merlot, 20% Syrah, 15% Cabernet Sauvignon, 7% Malbec, 4% Cabernet Franc and small amounts of Grenache, Montepulciano and Tempranillo. White grapes are predominantly Chardonnay and Viognier with Arneis, Gewurztraminer and Riesling featuring in small quantities.

http://www.gimblettgravels.com

Craggy Range Te Muna Vineyard.

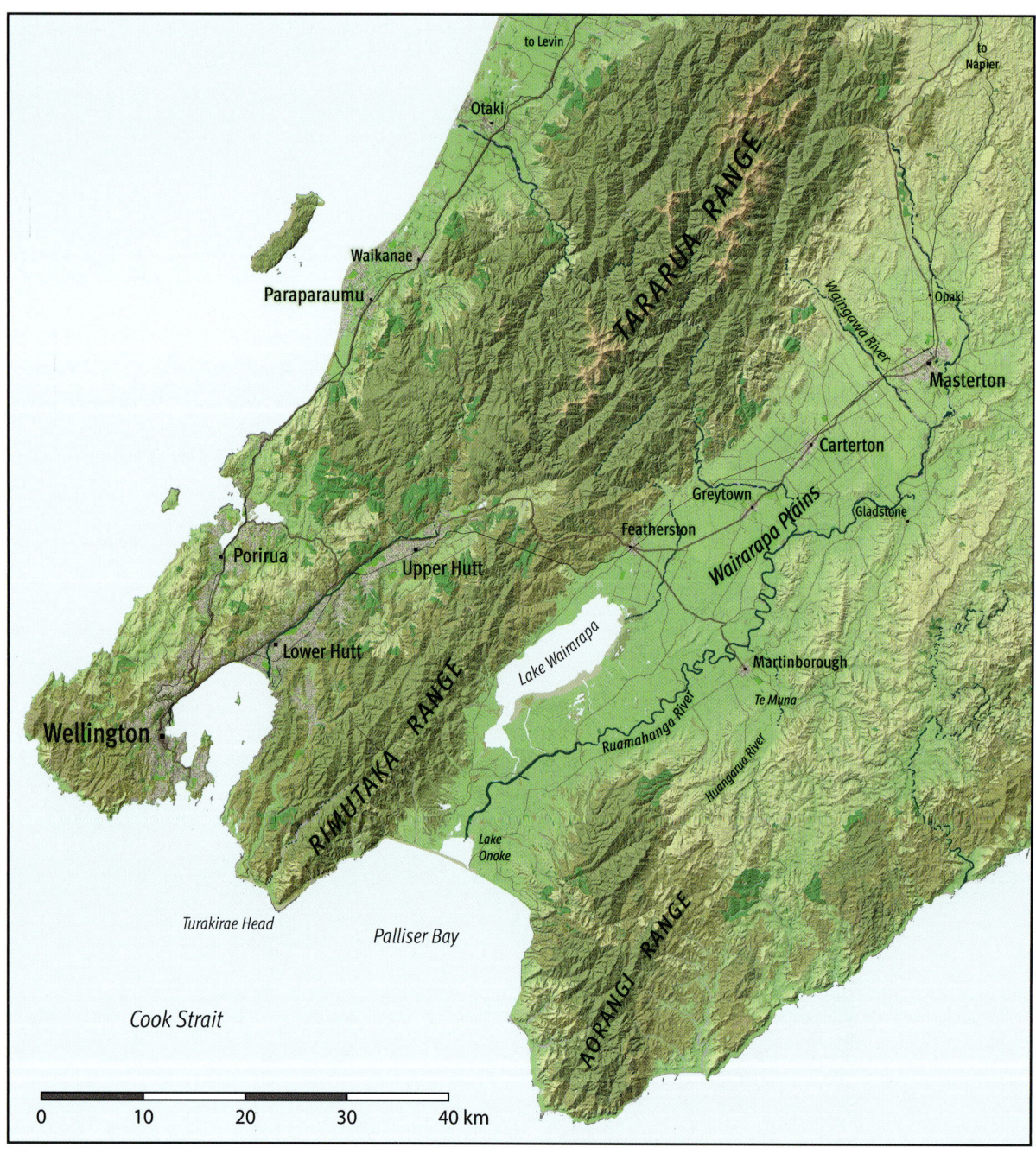

The Wairarapa region is located 30 kilometres north of New Zealand's capital city, Wellington and has built a reputation as a fine wine region based on the quality of its Pinot Noir.

In 1978, a soil science report prepared for the Government identified that Martinborough shared similar soils and climate to Burgundy. The discovery of an area with the combination of free-draining river gravels, similar temperatures and rainfall to Burgundy led to the foundation of this region.

Wairarapa stretches from southern Hawke's Bay down to the windy, south coast at Cape Palliser. The small township of Martinborough is the centre and its name is closely associated with the development of the region. Martinborough is roughly the same latitude as Picton, across Cook Strait, in the Marlborough Sounds.

Martinborough was the first wine area to prepare its designation under New Zealand's Geographical Indications (Wine and Spirits) Registration Act 2006, although the Act is still in the process of being implemented.

To the west, on the Kapiti Coast, is a new, small wine growing area of Ohau, located around the Ohau River that specialises in white wine.

HISTORY

Vines were first planted by William Beetham at Lansdowne Station, near Masterton in the 1880s. These included Pinot Noir and Syrah. When Masterton went 'dry' in 1905, during the height of the Temperance Movement, the vines were pulled out. Dr Derek Milne, one of the authors of the soil science report, was so convinced of the future of Martinborough as a wine region that, together with a group of five enthusiasts, he bought land to plant grapes. Neil and Dawn McCallum established Dry River in 1979 on land that had once been part of Dry

River Station. They were soon followed by Ata Rangi and Chifney (now called Margrain).

These vineyards are all located on what is now known as the Martinborough Terrace, an area along the Ruamahanga River consisting of dry, free-draining, alluvial river terraces.

CLIMATE

Wairarapa has the driest and coolest climate in the North Island. It is often described as having a semi-maritime climate based on its long coastline. The region is heavily influenced by its proximity to Cook Strait with strong winds and rainy weather coming from the south. The Tararua Range to the west, shelters the region from westerly rain while the Aorangi Range along the eastern coast protects it from this direction. Wairarapa is subject to regular frosts in winter and spring.

The Wairarapa has a high diurnal variation consisting of hot days and cool nights in summer and autumn which suits Pinot Noir as it is able to ripen slowly.

Late harvest and botrytised wines are produced in this climate.

SUB-REGIONS

The three main sub-regions of Wairarapa are Martinborough, Gladstone and Masterton, however wineries in Gladstone and Masterton prefer to describe their wine as 'Wairarapa'. They have a similar climate and produce Pinot Noir and Sauvignon Blanc as well as pockets of Pinot Gris, Riesling and Syrah.

Martinborough

Martinborough has a comparatively small area under vine, but it is highly regarded as being capable of producing elegant Pinot Noir, inspired by the grand crus of Burgundy, with their *'elusive delicacy'.* The dedication of the early Martinborough growers and winemakers has been fundamental in establishing New Zealand's international credentials as a producer of fine red wine beyond Sauvignon Blanc.

Alana Estate, Martinborough.

Te Muna is south east of Martinborough and located on the river terraces of the Huangarua River which flows into the Ruamahanga River near Martinborough. The alluvial river terraces are highly prized for the quality of their wines.

Climate: Cool spring, long hot, dry summer, windy.

Soil: Silt loam over free-draining gravels; limestone features in certain vineyard sites.

Predominant grapes: Pinot Noir, Sauvignon Blanc, Chardonnay, Riesling and Syrah.

Try these Pinot Noir: Ata Rangi, Alana Estate, Dry River, Palliser Estate, Martinborough Vineyards, Murdoch James Estate, Nga Waka, Te Kairanga, Kusuda.

Te Muna: Escarpment Vineyards, Craggy Range, Julicher.

Gladstone

Plantings of grapes started in Gladstone during the 1980s. This rural area, east of Carterton, picks up the important Ruamahanga River as it winds its way south towards Martinborough and the coast at Palliser Bay. The river terraces at Gladstone have a similar soil structure to Martinborough but with more room to expand. Dakin's Road has become the focus of plantings.

Climate: Cool climate with plenty of sunshine.

Soil: Some clay among the stony silt, free-draining river terraces.

Predominant grapes: Pinot Noir, Sauvignon Blanc, Riesling, Pinot Gris.

Try these wines: Urlar Pinot Noir, Schubert Pinot Noir Block B, Gladstone Vineyards, Borthwick Pinot Noir, Johner's selection of noble wines: Riesling, Sauvignon Blanc and Pinot Noir.

Masterton

Around Masterton, the largest town of the Wairarapa, there are a number of wineries both to the north and south. Opaki is a small area in the north again dominated by terraces formed by the Ruamahanga as it flows south.

Climate: Long, dry growing season with hot days and cool nights in late summer.

Soil: Free-draining, fine alluvial loams.

Predominant grapes: Pinot Noir, Sauvignon Blanc, Pinot Gris.

Try these wines: Matahiwi Pinot Noir, Wycroft Pinot Noir, Matahiwi Pinot Gris.

KEY WINE STYLES

Pinot Noir remains the hero of Wairarapa, made in an elegant style that is not as fruit driven as many from the South Island and with restrained use of oak. Sauvignon Blanc sits second, behind Pinot Noir, but it is perhaps the aromatics of Riesling and Pinot Gris that offer the best drinking.

Marlborough

Sauvignon Blanc vines on Scott Henry trellis, Dog Point Vineyard.

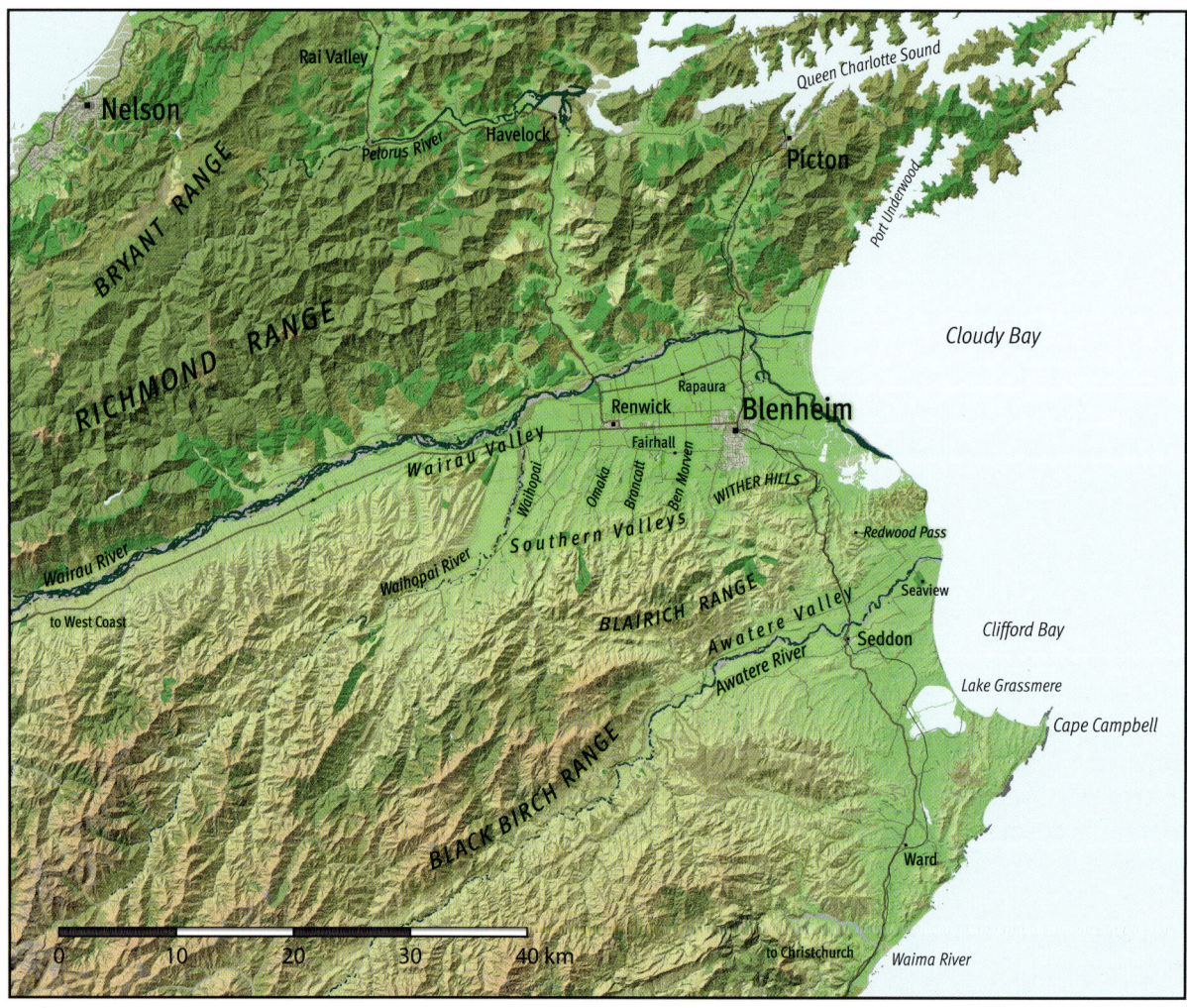

Marlborough is New Zealand's most important wine region. It has the greatest area of vineyards and Marlborough Sauvignon Blanc is not only recognised as a distinctive style of wine, but it alone has driven the surge of New Zealand wine exports since the 1990s.

There is something extraordinary about the combination of Marlborough's geography, geology and climate that makes is so well suited to being a leading wine region of the world. Many commentators believe that Marlborough's diurnal range consisting of hot days and cool nights allows the grapes to ripen slowly while retaining an intensity of fruit with high natural acidity.

The quality of Marlborough grapes has in turn attracted many new immigrants from overseas, often experienced winemakers, drawn to the exciting opportunities of this region. In the space of a few decades, thousands of hectares of farmland, primarily growing sheep and cattle or fruit trees, have been converted to vines.

HISTORY

The region has plantings across two main areas: the Wairau Valley and the Awatere Valley that lies to the south. Montana Wines planted the first commercial vineyard in August 1973. This was a mix of different grapes including Sauvignon Blanc, Muller-Thurgau and Cabernet Sauvignon. The first Montana Sauvignon Blanc was made from these vines in 1979.

Other wineries were soon to follow including Ross and Bill Spence from Matua who first experimented by growing Sauvignon Blanc in West Auckland. In 1983, Bill Spence and some other winemakers attended a wine conference in Perth and went on to Margaret River to taste wines. The group met David Hohnen of Cape Mentelle and Spence left behind some bottles of Marlborough Sauvignon Blanc. Hohnen commented, 'It was a bit sugary sweet, but the aromatics were amazing'. In 1984, Hohnen visited New Zealand to see for himself where these remarkable wines had come from. He also met Kevin Judd, who was then working for Selaks. Hohnen went back to Australia and raised AU$1 million to finance the winery, 'at a crazy interest rate of 23%'. Cloudy Bay's first vintage was in 1985 from 40 tons of purchased grapes with the wine made in Gisborne. In 1986, Judd constructed the winery and contracted Corbans to supply 120 tons of grapes per year. Cloudy Bay quickly began building its reputation. This early international investment in Marlborough, while small, was an important signal that people, external to the New Zealand wine industry, could see the potential of the region and grape variety and were prepared to take an enormous risk in a start-up region with no track record.

Other early winegrowers included Allen and Joyce Hogan from Te Whare Ra who planted grapes in 1979. Ernie Hunter arrived from Ireland and his wife, Jane, from Australia; they, along with Daniel Le Brun from France, were early innovators. Peter Vavasour, whose family had settled as farmers at the mouth of the Awatere River during the 1890s, planted grapes in 1985.

CLIMATE

Marlborough looks north towards Cook Strait, the stretch of water that separates the North and South islands. It can get very windy and the high hills on both islands create a funnel effect on the prevailing westerly wind. There is always wind in Marlborough. The heavily wooded Richmond Range, adjacent to Queen Charlotte Sound, traps rain from the west and

north-west and casts a rain shadow over the lower foothills, while the Wither Hills to the south and the Black Birch Range to their west shelter the region from exposure to the cold southerly rains.

The climate in Marlborough is considered temperate, maritime with high sunshine hours; however, there are significant climatic differences within the sub-regions. The Awatere Valley is considered cooler and drier than the Wairau and Southern Valleys.

Irrigation schemes utilising the major rivers provide new sources of water and allow the viticulturists to manage the supply of water, particularly in drought-prone areas.

SUB-REGIONS

There are three main sub-regions in Marlborough that form the basis of its Geographical Indication (GI). Within each sub-region, distinct districts have been further identified and show the complex character of Marlborough.

The Wairau Valley

The Wairau Valley is dominated by the Wairau River that flows from mountains, near the Nelson lakes, and pushes north-east through the flat plains towards the coast at Cloudy Bay. The Wairau's alluvial flood plain has many types of soils from light silt and sand to the greywacke river stones of Rapaura. The Wairau River is close to the Richmond Range and the vineyards grown along its banks, especially from Renwick to Rapaura, benefit from the light silt, sand, shingle and river stones. Many consider this area to be the premium grape-growing area of Marlborough. The soil changes again south and east of the river to heavier loams and clays.

Distinct areas: Rapaura, Lower Wairau, Conders Bend, Renwick, Kaituna.

Climate: Cool, dry and windy.

Soil: A great diversity of soils with the best considered to be young, stony soils of light loams (sand, silt and clay) over shingle, naturally free-draining.

Predominant grapes: Sauvignon Blanc, Riesling, Chardonnay, Pinot Noir, Gewurztraminer, Pinot Gris.

Yealands Estate, Awatere; looking south to Clifford Bay

Try these wines: Seresin Pinot Noir, Te Whare Ra Riesling, Wairau River Sauvignon Blanc, Hans Herzog Spirit of Marlborough (Merlot dominated Bordeaux blend), Allan Scott Sauvignon Blanc, Huia Pinot Gris, Hunter's Sauvignon Blanc, Isabel Chardonnay, Jackson Estate, Saint Clair Pinot Gris, Stoneleigh Rapaura Pinot Gris, Nautilus Sauvignon Blanc.

The Southern Valleys

The Southern Valleys consist of a series of valleys, south of Highway 63 and rising up towards the Blairich and Black Birch Ranges and Wither Hills towards the coast. The Fairhall and Brancott valleys are where Montana first planted grapes in 1973. The soils are older with heavier clay loams that hold more moisture than the shingles of Rapaura. The valleys of Waihopai and Omaka have rolling hills that become cooler as the altitude increases. Ben Morven is protected from southerly weather by the Wither

Hills, although this area is prone to drought. In the Southern Valleys, Pinot Noir is becoming the premium grape and the best vineyard sites are considered to be those planted on rolling hills with higher altitude.

Distinct areas: Ben Morven, Brancott, Fairhall, Omaka, Waihopai.

Climate: Cooler and drier.

Soil: The soils are much older, with clay loams in rolling hills that hold more soil moisture, along with weathered gravel.

Predominant grapes: Pinot Noir, Sauvignon Blanc, Chardonnay.

Try these wines: Clos Henri Pinot Noir, Brancott Special Reserve Fume Blanc, Fromm, Greywacke Sauvignon Blanc, Dog Point Chardonnay, Forrest Estate Sauvignon Blanc, Foxes Island Pinot Noir, Lawson's Dry Hills Pinot Noir, Wither Hills Sauvignon

Blanc, Charles Wiffen Chardonnay, Zephyr Riesling, Villa Maria Single Vineyard Southern Clays Pinot Noir (Ben Morven).

The Awatere Valley

The Awatere Valley is located further south and like the Wairau is dominated by its river, which runs from the mountains down to the coast at Clifford Bay. Flanked by the Black Birch Range of mountains to the north and the Inland Kaikoura Ranges to the south of this narrow river valley, there are many smaller valleys and rolling hills that break up the landscape. In less than 10 years, the Awatere has grown from a few experimental vineyards to having more area under vine than Hawke's Bay.

Awatere Valley is considered 'cooler, drier and windier' than the other sub-regions. Bud break is later than in the Wairau. The summers are long and dry, prone to drought, which makes water sourced from the Awatere River for irrigation an important asset. The diurnal change each day, with often a 10-degree drop in temperature overnight, extends the growing season.

The Awatere Valley has an exceptional terroir. It is considered to be geologically older than the Wairau and the dramatic river terraces of stony gravels and sandstone that have been carved out by the river offer unique individual vineyard sites. The risk of frost in spring and autumn, and drought, add to the challenges of growing grapes in this region.

Distinct areas: Blind River, Seaview, Seddon, Redwood Pass.

Climate: Overall it is cooler, but summer can get very hot, and there is more wind, less rain. Spring frosts occur and can also happen in late March around harvest time.

Soil: Clay on river plains, free-draining river terraces with alluvial gravels; sandstone.

Predominant grapes: Sauvignon Blanc, Pinot Noir.

Try these wines: Yealand's Sauvignon Blanc from Seaview, Vavasour Awatere Pinot Noir, Villa Maria Taylor's Pass Chardonnay.

Babich Wakefield Downs Vineyard, Awatere.

KEY WINE STYLES IN MARLBOROUGH

The predominant grape varieties in Marlborough are Sauvignon Blanc, Pinot Noir, Chardonnay, Pinot Gris and Riesling.

White wines, especially Sauvignon Blanc from the Wairau and Southern Valleys, are regarded as displaying the classic characteristic of Marlborough Sauvignon Blanc: intensely aromatic, both on the nose and on the palate; smelling like freshly cut grass; pineapple and passionfruit aromas; and high acidity. Awatere Valley Sauvignon Blanc is characterised by a distinct minerality, reflecting its cooler climate.

There are some excellent examples of Gewurztraminer grown in Marlborough. It retains those heady, fragrant aromas of lychees and rose petals. Chardonnay and Riesling also flourish as does Pinot Gris, which is a lighter, more neutral wine, but often has aromas of pears and nectarines.

Grapes for sparkling wine remain important and Marlborough is by far the biggest producer of this style of wine.

Pinot Noir is the rising star of Marlborough.

Nelson

Seifried Redwood Vineyard, Moutere Hills, Nelson.

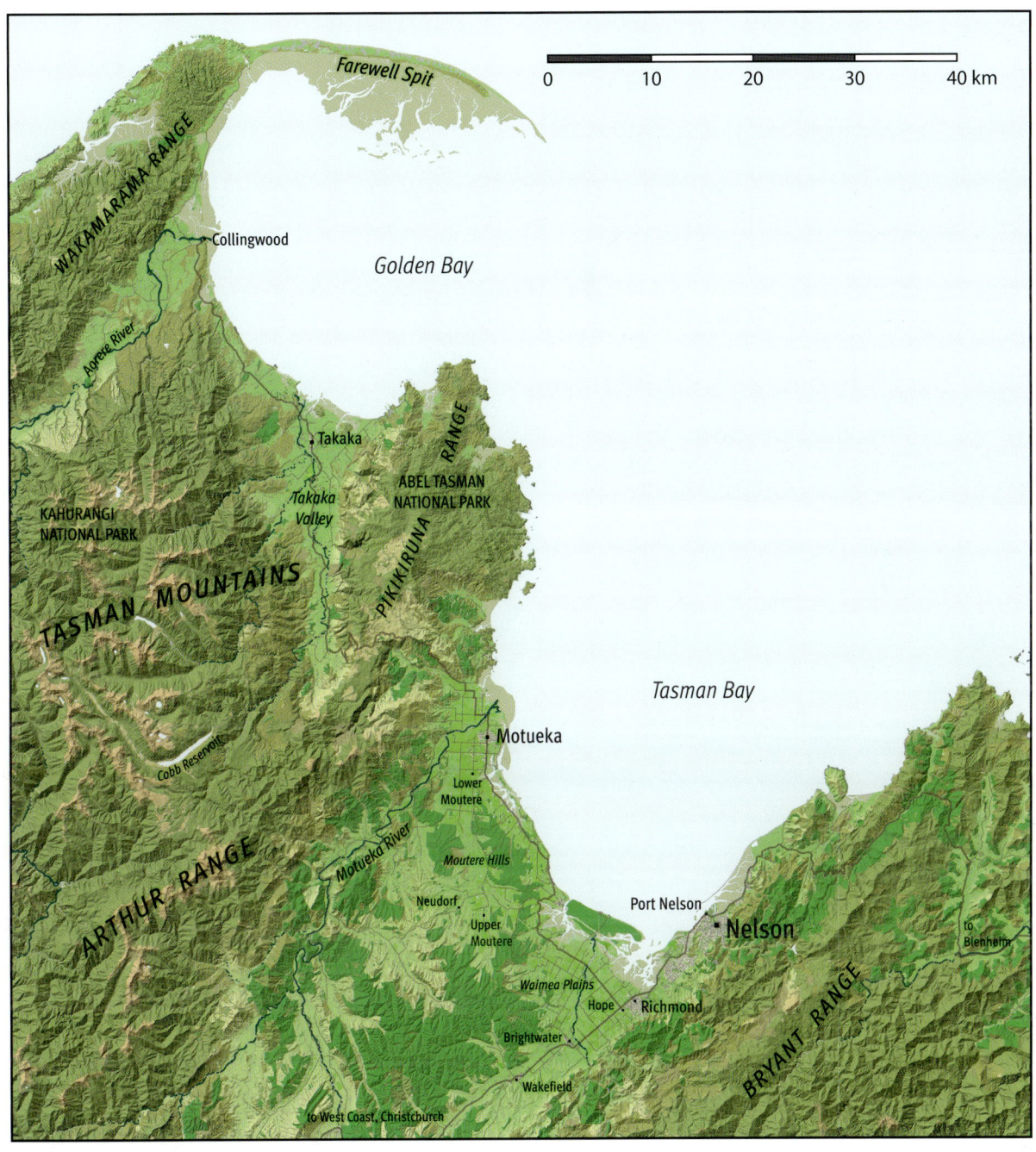

Nelson is situated at the top of the South Island, west of Marlborough. Nelson's three national parks, settled sunny weather and golden sandy beaches make it a very attractive visitor destination. While Sauvignon Blanc is the most planted grape, Nelson's wine reputation has, to a great extent, been forged by the fine wines of Judy and Tim Finn from Neudorf

Vineyards and the innovative aromatic wines of Agnes and Hermann Seifried.

Nelson has many small, family owned wineries. Vineyards are planted on rolling the Moutere Hills, west of Nelson City and in the Waimea Plains to the south of the city. The are also plantings along the coastal areas of Tasman Bay and on to Golden Bay. The Spencer Hill winery with vineyards in Moutere Hills and on the coast have developed a range of kosher wines under the label of Goose Bay.

HISTORY

Nelson was an early settlement of the New Zealand Company in 1842 and its temperate climate and fertile plains attracted many new immigrants. A small group of German settlers planted the first grapes in Neudorf (meaning 'new village') in the Moutere Hills in 1843, however they later left the region. Nelson developed a strong horticulture based economy with large fruit orchards, tobacco and hops. Many of the fruit orchards are now converted to grapes. Seifried first planted grapes in 1976 and Neudorf Vineyards established in 1978, in the same area as the German settlement.

CLIMATE

Nelson is flanked by mountain ranges to the east, south and west and therefore is protected from wet weather from the west and cold southerly rains from the south. The climate in Nelson is hot and dry in summer, with some of the highest sunshine hours in New Zealand. In summer, the daytime temperature can reach 30°C but afternoon sea breezes from Tasman Bay to the north, introduce a moderating influence.

SUB-REGIONS

There are two main sub-regions in Nelson where grapes are grown:

Moutere Hills

Upper Moutere, in the rolling Moutere Hills, offers some of the best Nelson wines.

Climate: Moderately hot and dry in summer.
Soil: Clay soils.
Predominant grapes: Chardonnay, Pinot Noir
Try these wines: Neudorf Moutere Pinot Noir, Neudorf Moutere Chardonnay, Woollaston Mahana Pinot Noir, Brackenbrook Gewurztraminer.

Waimea Plains

The Waimea Plains are south west of Nelson City and are part of the Wai-iti River flood plain. Hope is on the eastern side while Brightwater is further south. The majority of Nelson's grapes are grown in Waimea.

Climate: Moderately hot and dry in summer, maritime.
Soil: Gravelly silt loams, similar to Marlborough.
Predominant grapes: Sauvignon Blanc, Pinot Noir, Pinot Gris, Riesling, Chardonnay.
Try these wines: Waimea Sauvignon Blanc, Brightwater Vineyards Barrique Chardonnay, Greenhough Chardonnay, Woollaston Sauvignon Blanc, Seifried 'Sweet Agnes' Riesling with 165 g/L residual sugar.

KEY WINE STYLES

The main grape varieties grown are Sauvignon Blanc, Pinot Noir and Pinot Gris. Nelson is associated with the aromatic white wines produced from Pinot Gris and Riesling in dry and off dry styles. Gewurztraminer and Viognier are growing in popularity.

A handful of top quality Chardonnay and Pinot Noir are also important. Sparkling wines, made in the traditional method, are made by Kaimira and Woollaston.

Camshorn Vineyard, looking north across Waipara Valley, North Canterbury.

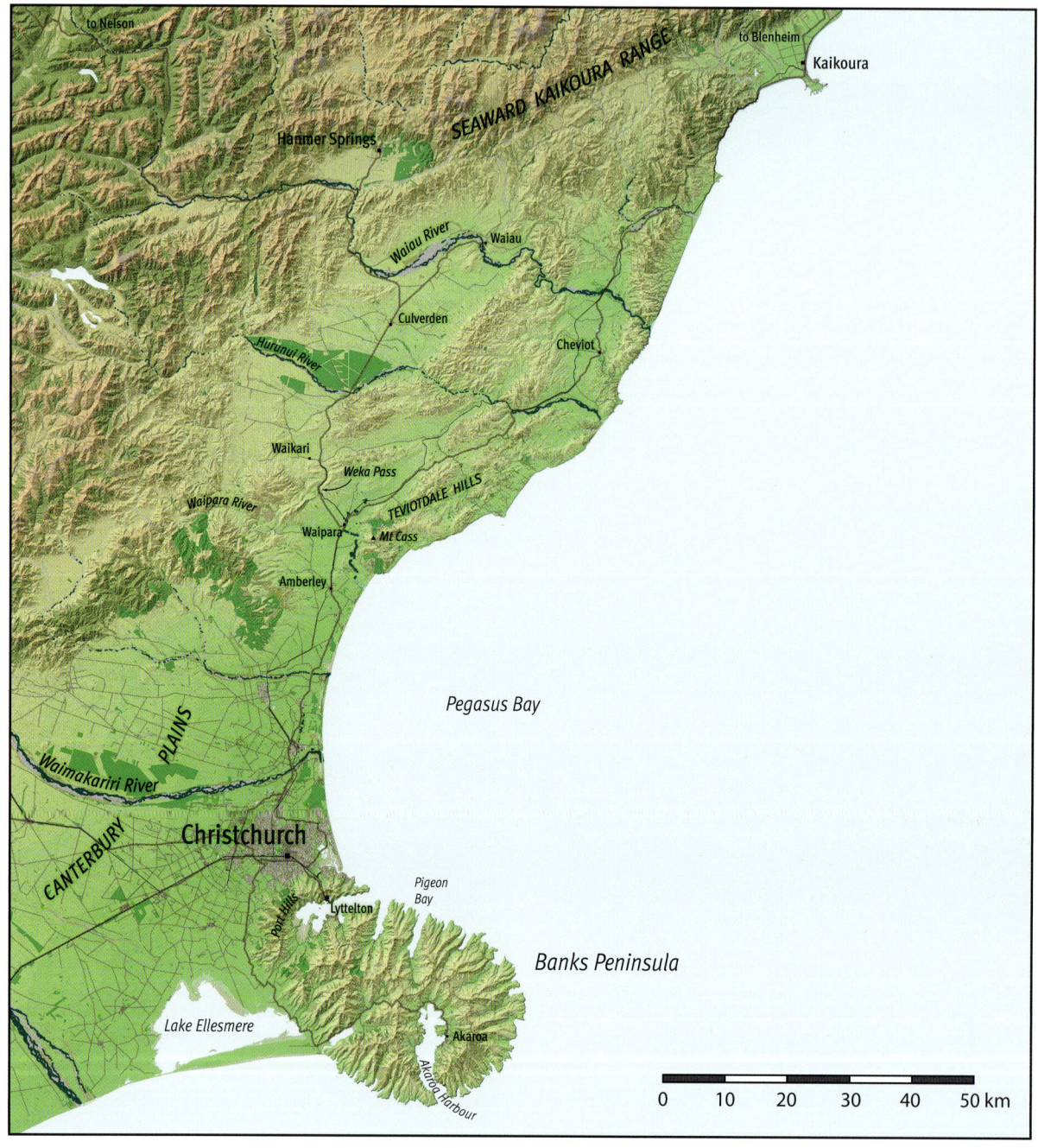

Canterbury is a large region with a great diversity of soils and microclimates. The Southern Alps dominate and lie to the west of the arable and extensive Canterbury Plains. A decade ago, most of the region's vineyards were planted on the outskirts of Christchurch City and describing the region as 'Canterbury' was appropriate. Today, most vineyards are 40 minutes' drive north in the Waipara Valley. This marks the preference to refer to the region as North Canterbury, however there still remain vineyards around Christchurch, on Banks Peninsula and in South Canterbury.

Lincoln University, near Christchurch, has played an important role in the establishment of New Zealand's modern wine industry. Lincoln started its first wine seminars in 1973 under the direction of Dr David Jackson and Danny Schuster. The ongoing innovation and teaching programmes from Lincoln have helped train many of New Zealand's leading viticulturists and winemakers.

Canterbury is my region and until 2011, Christchurch was our home and the location of the New Zealand School of Food and Wine and my restaurant, Hay's which focused on organic lamb from our family farm at Pigeon Bay. All that changed with the series of earthquakes that destroyed major areas of Christchurch including our home and the Hay's building in Victoria Street. The New Zealand School of Food and Wine is now based in Auckland's Viaduct but we still regularly return to our farm in Pigeon Bay.

HISTORY

Like Northland, Canterbury was one of the earliest regions to grow grapes. A group of French settlers arrived in Akaroa in 1840 and brought vine cuttings with them. These included Chasselas, a grape suitable for both eating and wine, however the vines did not go beyond a cottage industry that primarily supplied their own needs.

In 1976, Graham Steans, another academic from Lincoln, planted vines at Kaituna Valley on Banks Peninsula. He was encouraged by Danny Schuster to plant Pinot Noir. It was not until early 1978 that the first commercial vineyard was planted at St Helena on the Mundy brothers potato farm, north of Christchurch. They shot to fame when the St Helena Pinot Noir 1982, made by Danny Schuster, won a gold medal at the Air New Zealand Wine competition. Alex, Theo and Marcel Giesen, recently arrived from Germany, established their Burnham vineyard in 1981.

David Jackson continued with the Lincoln vine trials and tasting panels were established to review the wines. Over the years, the panels included many luminary figures; from the medical profession: Professor Ivan Donaldson, who went on to found

Pegasus Bay Winery with his wife, Chris and Professor Don Beaven. Other members included Hermann Seifried, Geoffrey Collard and Neil McCallum. Enthusiasts also attended the early seminars at Lincoln including Tim Finn, Ernie Hunter, Rolfe Mills and Ann Pinckney from Central Otago.

Canterbury is currently the fifth largest wine region in New Zealand.

CLIMATE

The climate in Canterbury varies considerably from North Canterbury and the Waipara Valley to that around Christchurch. Both areas are protected from rain by the Southern Alps, although the blustery nor'west winds can quickly throw Canterbury into drought. North Canterbury is also sheltered from the cold southerly weather by the Teviotdale Hills on the coast. This area is drier and gets very hot in mid-summer with southerly rains hitting the coast and not the inland areas. This makes irrigation essential. All of Canterbury is frost-prone, depending on the specific location of the vineyard.

SUB-REGIONS

North Canterbury/Waipara

The Waipara Valley is approximately 5 kilometres from the coast and around 80 metres above sea level. Many of the vineyards are planted on free draining gravels soils on terraces of the Waipara River. Vineyards planted on the rolling hills consist of clay and fragmented limestone.

New areas have developed north west of Waipara at Waikari, near Weka Pass, where Bell Hill and Pyramid Valley vineyards are located; these have high limestone and clay soils. Both vineyards have established an international following for their wines, especially Pinot Noir and Chardonnay. Waikari is nearly 40 kilometres inland and about 200 metres above sea level and the vines are often harvested at the end of April.

Climate: Hot days and cooler nights in summer and autumn, spring frosts.

Soil: Free-draining gravels soils as well as clay, limestone.

Predominant grapes: Pinot Noir, Sauvignon Blanc, Riesling.

Try these wines: Alan McCorkindale Riesling, Black Estate Chardonnay, Mountford Pinot Noir, Pegasus Bay Prima Donna (Pinot Noir) and Virtuoso (Chardonnay), Bellbird Springs Sauvignon Blanc, Greystone Pinot Noir. Terrace Edge 'Liquid Geography' Riesling.

Other areas

Mount Beautiful vineyard is planted on the southern terraces of the Wairau River, north of Cheviot in North Canterbury. This new development is dominated by Sauvignon Blanc and Pinot Noir plantings.

KEY WINE STYLES

There are a number of different grape varieties grown in Canterbury but the two most important are Pinot Noir and Riesling, although Sauvignon Blanc is the most planted vine.

Pegasus Bay Winery's Riesling followed by the late harvest and botrytis influenced Aria and Bel Canto Rieslings have provided the region with excellent models that others vineyards are now emulating.

Pinot Noir remains the other important grape from Canterbury with wines made in a variety of styles from lighter, summer fruits of raspberry and strawberries to full-bodied black fruits with savoury characters. A small number of premium Chardonnay is also produced.

Sauvignon Blanc is widely planted with Chardonnay, now overtaken by the Pinot Gris.

Central Otago

Felton Road Vineyard, Bannockburn, Central Otago.

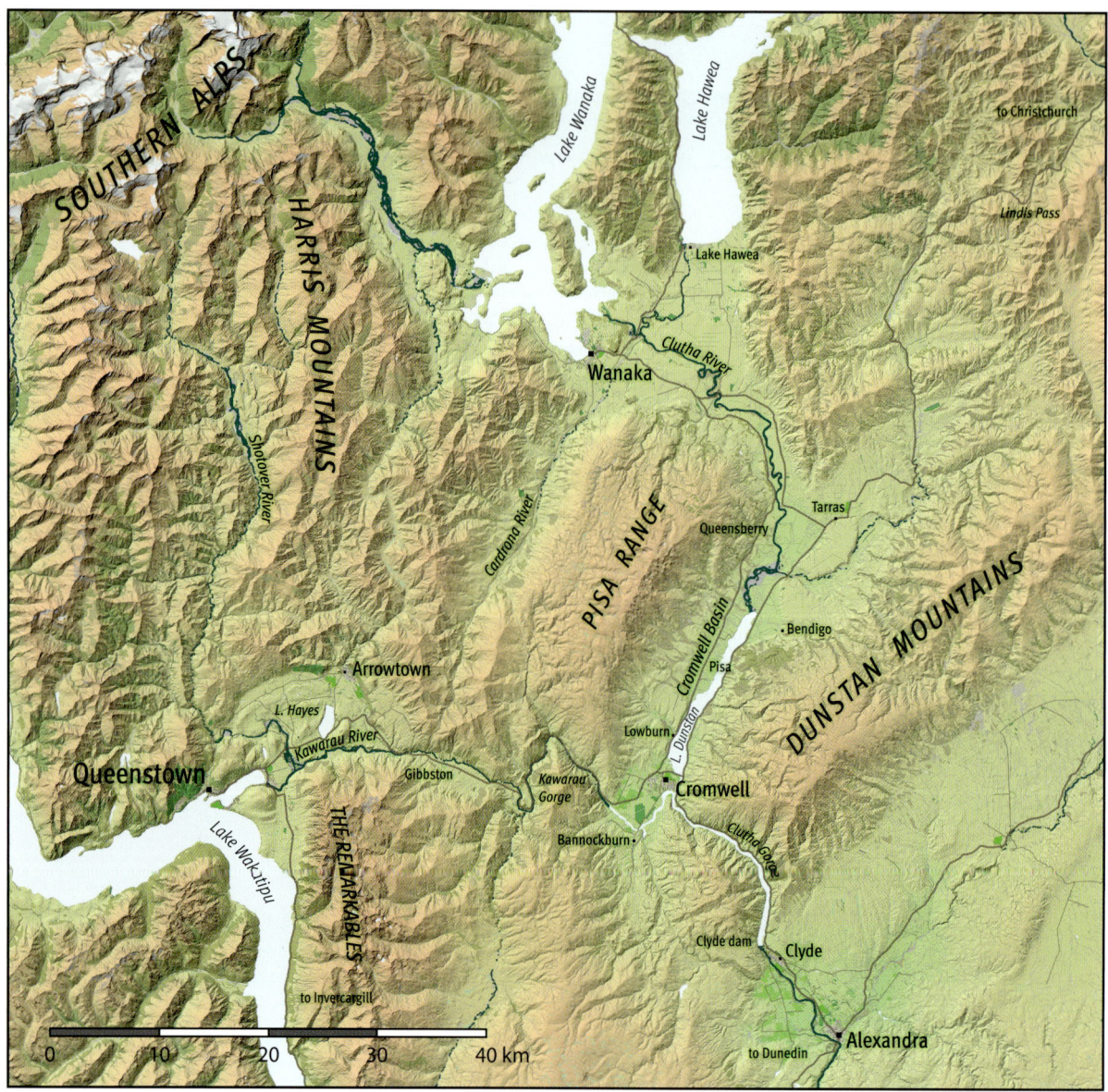

Central Otago is New Zealand's premier international tourist destination. It continues to capture the imagination of visitors as they stand spell-bound by the dramatic landscape of river-valleys, crystal-clear lakes, towering mountains and brilliant light. Equally, the international reputation of the wines produced from this region has grown with dazzling speed.

Central Otago refers to the inland areas of the historic province of Otago with Dunedin as its city. It stretches from the Waitaki River where the province bounds Canterbury, west towards the Southern Alps and includes the spectacular lakes of Wanaka and Wakatipu.

The two major rivers, the Clutha from Lake Wanaka and the Kawarau from Lake Wakatipu, join at Cromwell when the Clyde Dam captures their water to generate electricity before the Clutha heads out to the coast, south of Balclutha. The new region of Waitaki, in North Otago, offers a completely unique growing environment.

Central Otago has built its reputation as a wine region with Pinot Noir which currently makes up 80% of its vines. The wine-growing areas that have been planted since the 1990s, have created a number of sub-regions with their own distinct characteristics. Climate and soil structure remain the dominant influences in the different styles of wine that are produced.

From the edge of Lake Wanaka to the Cromwell Basin, heading east to Alexandra or west through the Kawarau Gorge to Gibbston Valley, there are numerous vineyards primarily growing Pinot Noir. White grapes, particularly Riesling, Pinot Gris and Chardonnay are however, also producing outstanding wines and this confirms that Central Otago is not only about Pinot Noir. Central Otago Sauvignon Blanc also has its own personality.

Apart from Patagonia in Argentina, Central Otago is the world's most southern wine-growing region.

HISTORY

Grapes were first planted in Central Otago in 1864 by a Frenchman, Jean Desire Feraud, who won a prize for his New Zealand 'Burgundy' in a Sydney wine show in 1881. It was not, however, until the 1980s that a small group of pioneers began growing grapes and making wine.

Rolfe and Lois Mills planted grapes at Rippon Vineyard on the shores of Lake Wanaka in 1982. Bill Grant planted grapes in Alexandra as did Verdun Burgess and Sue Edwards; Ann Pinckney planted vines at Dalefield. Alan Brady established Gibbston Valley Wines, now one of the region's biggest wineries, which released its first commercial wine in 1987. In the same year, Rob and Greg Hay planted vines at Chard Farm.

CLIMATE

At a latitude of 45 degrees south, surrounded by large mountain ranges, Central Otago is considered to have New Zealand's only continental climate without the moderating influence of the sea. The Southern Alps cast an important rain shadow as the wet, westerly weather drops its rain in the mountains.

As a consequence, rainfall is surprisingly low. In summer, temperatures can regularly climb above 30 degrees Celsius with still, balmy evenings. As autumn approaches, there can be a large diurnal temperature range with hot days in the mid 20s and then cooling down at night time to below eight degrees.

In winter, heavy frosts are a regular occurrence with some vineyards experiencing frequent falls of snow. The wind can be bitterly cold. Frost is a risk both in spring, during flowering, and in autumn as harvest approaches.

SOIL

The soil structure of Central Otago differs considerably from other New Zealand wine-growing regions. The river valleys are formed by ancient glaciers which have left heavy deposits of silica, quartz and mica. These are often referred to as loess. The crystals of quartz, mica and other minerals mix with the river gravels to form free-draining soils which grow vines with low vigour.

For winemakers, loess is considered very desirable for Pinot Noir.

SUB-REGIONS

There are differences in climate and soil between the sub-regions in Central Otago and this translates through to a variety of wine styles. Some very successful regional brands are emerging to capture the everyday drinking market. These include Mt Difficulty's Roaring Meg with Central Otago Sauvignon Blanc, Pinot Gris, Riesling and Pinot Noir. The very successful Rabbit Ranch wines include Central Otago Pinot Noir and Pinot Gris with Sauvignon Blanc sourced from Marlborough.

Gibbston

Gibbston, in the Kawarau Gorge, is considered to have the coolest climate in Central Otago. Pinot Noir, grown on these slopes, has naturally higher acid levels often with lighter, fruit flavours such as red cherries, strawberries and red currants. Many of the Gibbston vineyards have chosen to expand their production by sourcing grapes from the Cromwell Basin area.

Climate: Long, cool, dry summer.

Soil: Free-draining river gravels, quartz, mica.

Predominant grapes: Pinot Noir, Riesling.

Try these wines: Mount Edward, Valli, Peregrine, Chard Farm, Amisfield, Gibbston Valley, Kalex.

Bannockburn

Bannockburn is south of Cromwell township on the north-facing Kawarau River terraces. From the 1860s, for over fifty years, the Bannockburn sluicings attracted gold miners in search of their fortunes. The same soil structure now attracts grape growers. The soil is primarily free-draining sandy loams with schist gravels. Bannockburn is considered one of the warmest and driest areas of Central Otago.

Climate: Long, warm, dry summer.

Soil: Free-draining sandy loams with schist gravels.

Predominant grapes: Pinot Noir, Riesling, Chardonnay.

Try these wines: Felton Road Block 5, Felton Road Block 2 Chardonnay, Mt Difficulty Target Gully Pinot Noir, Terra Sancta Pinot Noir, Mt Difficulty Long Gully Chenin Blanc, Carrick Unravelled Pinot Noir, Akarua Pinot Gris, Ceres Pinot Noir.

Cromwell Basin

The Cromwell Basin is the area north of the township of Cromwell which is adjacent to Lake Dunstan. The regions on the western side include Lowburn and Pisa and Bendigo, on the eastern side of Lake Dunstan.

Cromwell

This area is near the township of Cromwell.

Try these wines: Wooing Tree, Northburn Station, Rockburn, Serendipity.

Lowburn

Lowburn and Pisa lie on the east-facing foothills of the Pisa Range of mountains. The attract all day sun and have become a desirable area for new plantings. Many of the Gibbston wineries also have vines planted here.

Try these Lowburn wines: Aurum, Mount Michael, Burn Cottage, Maude.

Try these Pisa wines: Ellero, Locharburn, Surveyor Thomson, Charcoal Gully Sally's Pinch Pinot Noir.

Bendigo

Bendigo faces north-west near the junction of Cromwell and the Lindis Valleys and is developing a reputation for aromatic white wines.

Try these wines: Mondillo Riesling, Prophets Rock Pinot Noir, Misha's Gewurztraminer, Misha's High Note Pinot Noir, Aurora Riesling, Quartz Reef Pinot Gris.

Alexandra

Alexandra lies further south along the Clutha River. It is perhaps the region with the greatest diurnal range particularly during autumn.

Try these wines: Judge Rock, Shaky Bridge, Grasshopper Rock, Drumsara, Two Paddocks.

Wanaka

The Mills Family of Rippon Vineyard consistently demonstrate that the unique terroir of Lake Wanaka can produce outstanding wines.

Try these wines: Rippon Pinot Noir comes in a number of forms as the Mature Vine, Emma's Block and 'Jeunesse' from the younger vines. The Maude Chardonnay, which has been hand harvested, foot stomped and basket pressed then fermented in seasoned French oak puncheons.

Waitaki

The Waitaki region is Inland from the historic town of Oamaru, famed for its limestone architecture. This small wine region has been planted on limestone-rich soils. Harvest does not usually occur until late April, early May. The white grapes of Pinot Gris, Riesling and Gewurztraminer show a flinty minerality while the Pinot Noir from Waitaki can be elegantly restrained through to full-bodied.

Climate: Temperate summer with sea breezes from the coast, long, warm, dry summer and autumn.

Soil: Limestone, alluvial gravels.

Try these wines: Ostler, Valli, Pasquale, Waitaki Braids

The Rippon Vineyard, Lake Wanaka, Central Otago.

KEY WINE STYLES

Pinot Noir remains the most important grape but as the vines age, there is more diversity of styles emerging. With such a cool climate, white wines flourish and the emergence of aromatic Gewurztraminer, Pinot Gris, Riesling and even

Albarino is raising the profile of Central Otago's white wines.

Felton Road's Chardonnay is proof that top quality Chardonnay can be produced.

New Zealand Winegrowers

The extraordinary growth of the New Zealand wine industry would not have been possible without the commitment and dedication of large numbers of contract grape growers. In Marlborough, many of these growers were pastoral farmers or pip fruit growers looking to diversify and add new revenue streams. Many of these conract grape growers have gone to develop their own brands and labels.

New Zealand Winegrowers is the national organisation for New Zealand's grape and wine sector. It was established in 2002 as a joint initiative of the New Zealand Grape Growers Council (independent or contract grape-growers) and the Wine Institute of New Zealand, which acts for all New Zealand wineries.

Winegrowers' role includes:

- Advocacy at regional, local and international levels.
- Global marketing coordination.
- Facilitation of research on industry priorities.
- Giving the industry timely and strategic information.
- Organising events including Bragato Conference and Awards and the Air New Zealand Wine Awards.

The organisation is primarily funded through a levy on the sale of grapes collected by the Grape Growers Council and a levy on the sale of wine collected by the Wine Institute under the Wine Act 2003.

Please look at the New Zealand Winegrowers website, which holds a huge amount of information: http://www.nzwine.com

WINE PRODUCER CATEGORIES

As the wine industry has grown, New Zealand Winegrowers has started to define wineries by their production levels. There are three tiers, with the major producers in Category 3. Wine exports are the driver of this growth, with Marlborough Sauvignon Blanc the flagship wine style.

Category 1: annual sales not exceeding 200,000 litres

In 2013, there were over 600 wineries at this level. This includes boutique wineries such as De La Terre in Hawke's Bay, Burn Cottage in Central Otago and Tongue in Groove in North Canterbury. While Category 1 wineries are considered small, many still exert a significant influence over the New Zealand wine industry and have international reputations. Winemakers include Blair Walter from Felton Road, Kevin Judd from Greywacke (his own label; he was also founding winemaker of Cloudy Bay), Warren Gibson from Trinity Hill and Bilancia, Mat Dicey of Mt Difficulty, Tim Finn of Neudorf and Clive Patton from Ata Rangi.

Category 2: annual sales between 200,000 and 4,000,000 litres

In 2013, there were 75 wineries at this level. This includes wineries such as Pegasus Bay, Spy Valley, Tohu Wines, Sacred Hill, Te Mata, Saint Clair, and Seifried.

Category 3: annual sales exceeding 4,000,000 litres

In 2013, there were 10 wineries at level. This category is for the largest wineries and includes Villa Maria, founded and owned by Sir George Fistonich and his family. Their brands include the Villa Maria range of wines as well as Esk Valley, Vidal, Thornbury, and Te Awa.

Listed winemaker Delegat's owns the very successful Oyster Bay brand. Peter Yealands established his own winery in 2008 based at Seaview at the mouth of the Awatere River with over eight hundred hectares of vines and is focused on growing export markets with a number of brands including Crossroads from Hawke's Bay.

International investment

Investment from overseas has been an important source of capital for the New Zealand wine industry. The process of buying land, establishing vines, building winemaking facilities followed by international marketing is expensive and many wineries have found that they needed significant investment to turn their dreams into reality.

Cloudy Bay, considered an iconic New Zealand brand, was founded by David Hohnen from Margaret River in Western Australia who sold shares to Veuve Clicquot around 1990. In 2000, Hohnen sold his remaining shares. Veuve Clicquot was later acquired by LVMH.

Montana Wines, the leading producer at the beginning of the twenty-first century, was acquired by British listed company Allied Domecq in 2001. Its brands included Montana (now Brancott), Lindauer, Corbans, and Church Road. Its French rival, listed company Pernod Ricard, acquired Allied Domecq in 2005. Currently, Pernod Ricard's New Zealand investments include Brancott (formerly Montana), Church Road, Stoneleigh, Triplebank, Camshorn, Five Flax, and Deutz. It also owns and distributes many international spirit brands.

New Zealander Steve Smith MW and Australian investor Terence Peabody own Craggy Range. Craggy Range has expanded beyond Hawke's Bay to establish vineyards in Martinborough, Central Otago and the Awatere Valley.

Matua Valley, founded by Ross Spence in 1969 who was later joined by his brother, Bill, is now owned by Treasury Wine Estates, a listed Australian-based global winemaking and distribution business that was earlier part of the Fosters Group and includes Australian flagship winery Penfolds. Matua owns the Shingle Peak brand. Lion was formed in 2009 when the Japanese brewer Kirin purchased Lion Nathan. Lion is Australian based and owns New Zealand brands Wither Hills, Lindauer, Corbans, Daniel Le Brun and Huntaway. Like many of the corporates, it has its own highly effective sales and distribution system.

American public company Constellation owns Nobilo, Kim Crawford, Drylands, Selaks and Monkey Bay. The giant American winemaker, E&J Gallo is associated with Whitehaven. The Portuguese company Sogrape, who have built Mateus Rosé into a global brand since 1942, owns Framingham. In 2003, American investor Julian Robertson purchased Dry River and later, Te Awa Winery (recently sold to Villa Maria). The Foley Family vineyards from California own Clifford Bay, Vavasour and Te Kairanga with other acquisitions planned.

Australian company Accolade Wines purchased Mud House, Waipara Hills, Dusky Sounds, Haymaker and Skyleaf brands. The 400 hectares of vineyards owned by Mudhouse were sold to Hong Kong-based CK Life Sciences. They have leased the vineyards back to Accolade.

NOTES

1. Interview with Michael Brajkovich MW; quote in Thomson, p. 79.
2. Tyack, K. (2012), p. 52.
3. Thomson, J. (2012), p. 68.

CHAPTER TWO:

Growing grapes

Bud burst.

Viticulture is the science of growing grapes to produce wine. There are many factors that influence the taste of a grape and this starts in the vineyard.

New Zealand has a temperate, maritime climate which directly influences the grapes that can be grown. A temperate climate means that the seasons are not extreme, with rainfall spread throughout the year and sea breezes moderating the heat. In general the temperatures are cool to moderate with summers not very hot and winters that are not very cold.

The diurnal or daily range measures the temperature variation between day and night. This means, for instance, in summer that the daytime temperature can get to 25°C and then drop to a cool 10°C in the evening. The cool nights enable the ripening grapes to rest until morning when the temperatures heat up again. As a result, the New Zealand growing season is often up to three weeks longer than in Europe. The longer hanging time results in the development of more intense fruit flavours and the retention of higher levels of acid.

New Zealand's very high levels of light intensity with higher levels of ultraviolet light may also have a major influence on why the grapes develop pronounced, upfront, fruit-forward characteristics. Grape varieties that flourish in cool, temperate climates include Sauvignon Blanc, Pinot Noir and Riesling.

Vitis vinifera

Most grapes used for wine production come from the species of grape known as *Vitis vinifera*. This is the classic species native to Europe, the Mediterranean and south-western Asia. There are thousands of different varieties of this species, although fewer than 100 *Vitis vinifera* grapes are important to winemaking. Winemakers in countries such as Italy and Greece are rediscovering old, forgotten grape varieties and reviving them, but these grapes are considered novel rather than mainstream.

The main grape varieties used to make wine have been favoured because of their ability to grow healthy fruit with good flavours. Other important attributes are the ability to ripen, retain acidity and develop tannins for red wine production.

Different varieties of grapes need different conditions to grow well. The latitude, altitude, climate, weather, soil and aspect all play a role. During photosynthesis, grapes convert the energy of sunlight into chemical energy that can later be used to fuel the plant's growth. This chemical energy is stored in sugars,

created from carbon dioxide and water. This process results in the desirable flavours and characteristics required for winemaking.

It is also good to remember that the grape plant is a vine and therefore instinctively wants to grow prolifically, sending out shoots and searching for something to climb on to. This is why vines need to be trained and their growth managed so that the energy goes into growing fruit rather than more branches and leaves.

POLLINATION AND FRUIT

Most *Vitis vinifera* vines are hermaphroditic, which means they are able to self-pollinate as each flower has both male stamens and female ovaries. During the flowering of the vine, pollen is released and fertilises the ovary. The fertilised ovary then, over time, grows to take the shape of a grape berry. While wind and insects can aid pollination, the process can be self-contained within one vine.

CROSSINGS

A crossing occurs when a new grape variety is created following the cross-pollination of two vines of the same *Vitis vinifera* species. This can happen in two ways, as follows.

Natural crosses

Historically, vines crossed naturally in the vineyard during flowering. It is believed that Pinot and Gouais Blanc (an ancient grape variety not currently commercially grown) have given birth to at least 21 grape varieties including Chardonnay, Aligoté and Gamay Noir. Cabernet Sauvignon is also the result of a crossing of Sauvignon Blanc with Cabernet Franc.

Nursery pollination

This occurs when vines are deliberated crossed in the nursery. Muller-Thurgau was created in this way in 1882. It was a cross between Riesling and Madeleine Royal. Muller-Thurgau was also used to breed other crosses such as the high-yielding, early-ripening vine Reichensteiner, which is still grown in New Zealand.

Crosses sometimes result in a mutation – an accidental change in the DNA that results in different characteristics, such as the colour of the berry. Pinot Gris and Pinot Blanc are now recognised as colour mutations of Pinot Noir as they show the same DNA profile.

CLONAL SELECTION

Seeds of grapes are not generally used for commercial vine propagation, as viticulturists believe their progeny is unreliable. Far better results are achieved by taking a woody cutting from an existing healthy vine, known to produce quality fruit. This is often referred to as asexual reproduction or vegetative propagation.

The new vine is a product of only one parent; therefore it will be an exact genetic copy or 'clone'. Cuttings are selected based on the desirability of their genes. The cutting is planted in soil so that it takes root naturally.

Winegrowers often talk about using different clones of the same grape variety. They believe that this diversity produces a more complex wine. New Zealand Pinot Noir is often the result of a selection of Pinot Noir clones that are grown for different attributes.

The most famous clone grown in New Zealand must be the Abel or Ata Rangi clone. Clive Paton tells the story of how his friend Malcolm Abel, a part-time winemaker and New Zealand Customs officer, was working at Auckland Airport during the mid 1970s and had confiscated a single vine cutting, allegedly taken from Burgundy's finest estate, Domaine de la Romanée-Conti. The illegal cutting was then sent to the government's viticulture research station and Malcolm waited patiently for the first cuttings to become available and duly planted them.

This clone remains at the heart of Ata Rangi Pinot Noir. *'We love the texture and length of palate it delivers. Its tannins are substantial, yet are incredibly silky and fine. From our site it brings dark cherry, and a brooding, savoury feel.'* [1]

The original Sauvignon Blanc plantings in Marlborough were based on a clone from the University of California (Davis), called TK05196, which Ross Spence had identified in the Corbans' trial block in Kumeu in 1970. The trial block was destroyed shortly afterwards.

An alternative method is to graft the cutting, now known as the scion, on to another rootstock already established in the ground. This is known as top grafting. (See page 56).

HYBRIDS

A hybrid occurs when a new grape variety is created following the cross-pollination of two different vine species, e.g. *Vitis vinifera* x *Vitis rupestris*.

Hybrids were originally introduced to help counteract the impact of phylloxera and other vine pests and diseases. Seibel and Baco (Blanc) are examples of hybrids planted in New Zealand before the growth in reputation of varietal wines during the 1960s. Baco (Blanc) is one of the few hybrids permitted in the PDO category for European wines and remains an important grape in Armagnac.

VINE AGE

During the first two to three years after planting, the vine is in the establishment phase and not expected to bear fruit for wine. The focus is on building a root system in the soil. The root then searches out water and nutrients to feed it and ensure the development of a strong trunk for future harvests.

While young vines become productive after this time, the experts believe that a vine starts to reach maturity after about six to eight years, producing grapes with more flavour. The vine can go on to be productive for up to 40 years or even longer, although yields may start to decline after 25 years. As the vines mature, they should produce better fruit with more concentration.

VINE PROPAGATION AT RIVERSUN NURSERY, GISBORNE

1. Preparing the scion wood for grafting.

2. Grafting.

3. Taping the graft.

4. Grafted vines are packed into pumice.

5. Grafted vines begin to grow in pumice.

6. Grafted, potted vines in the nursery.

7. Potted vines.

8. Grafted field-grown plants that will be lifted and supplied as dormant vines.

9. Bundles of grafted vines ready to send to vineyards.

TOP GRAFTING

1. An incision is made in the receiving trunk.

2. The scion is attached.

3. The scion is taped to hold it firmly in place.

4. Grafted shoot is now growing from rootstock.

Phylloxera-infected vines.

Impact of phylloxera

Phylloxera is an aphid-like, sap-sucking insect that feeds off the vine's roots and leaves. This results in deformations on roots and secondary fungal infections which can strangle roots, gradually cutting off the flow of nutrients and water to the vine. The damage is usually not visible until the vine is dying. Phylloxera can easily spread to other vines and vineyards by wind, machinery and on shoes. It was first discovered in 1863 in southern England and in 1866 on vines in the Rhône Valley and Languedoc. Phylloxera is native to North America and was introduced to Europe on imported cuttings.

Romeo Bragato identified phylloxera-infected vines when he visited New Zealand in 1895. He had seen the damage caused to European vineyards and was anxious to prevent this happening in New Zealand. Phylloxera has had a devastating effect on vineyards, requiring replanting of mature vines and hence loss of production. For more than a century, phylloxera has forced viticulturists to experiment with ways to resist its spread.

Phylloxera infestation led to the development of hybrids where *Vitis vinifera* varieties were crossed with various North American vines that have a natural resistance to phylloxera. These included *Vitis rupestris, Vitis riparia, Vitis berlandieri* and *Vitis labrusca.* Over time, it was discovered that grafting on to American rootstock would offer the best solution.

Growth cycle of a grape

1. Bud break
Small buds burst out of previously dormant wood.

2. Flowering
The fertilisation of the cluster.

3. Fruit set
When flowers have been fertilised and the fruit for the harvest has been determined.

4. Veraison
Red grapes change from green to red. Sugar levels begin to rise as acidity levels drop. This image is of Pinot Noir.

BUD BREAK

In spring, the growth cycle begins with bud break, also called bud burst. The timing of bud break depends on the season. It usually starts in September when the days warm up to over 10°C. Generally the warmer regions such as Auckland will experience this first. Within 10 days of bud breaking, young leaves and minute clusters of flowers appear on the growing shoots.

FLOWERING

The tiny flower clusters, which actually resemble tiny grape bunches, develop into larger individual florets. The cap covering the floret falls off, exposing the flower parts (anthers and ovary). In late spring and early summer, up to 13 weeks after bud break, flowering begins.

FRUIT SET

The small green flowers self-pollinate and when fertilisation has occurred the fruit is said to be 'set'. The anthers of the flower drop off and the tiny hard, green berries resemble miniature bunches of grapes.

Coulure occurs when there is an imbalance of carbohydrate levels in the vine tissues and some berries fail to set or just fall off the bunch. This is caused by soil or weather conditions (cool, cloudy, wet) and it will reduce the yield. Some varieties of grapes, such as Merlot and Malbec, are more susceptible to coulure. Millerandage occurs during fruit set when different-sized berries, some without seeds, are formed (hen and chickens) within the same bunch. It is a result of poor pollination and is often caused by bad weather during flowering and fruit set. It is often characteristic of Pinot Noir and the Mendoza clone of Chardonnay.

VERAISON

The berries grow rapidly, but when they have reached about half to two-thirds their final size, growth temporarily slows. Veraison is the beginning of the ripening process, as sugars begin accumulating and acid levels drop. The fruit begins to soften and its colour changes, taking on their characteristic hues. After veraison, the fruit ripens, becoming sweeter, with reduced acid levels, and accumulating more colours, flavours and aromas. The grapes are now ready for harvest.

SEASONAL VARIATIONS

The goal of the viticulturist is to get healthy grapes to ripen so that they can be harvested about 100–110 days after flowering. Seasons, however, can vary from one to another in many ways. The weather is the major influence on this process.

Starting with bud break, cold, rainy, windy weather can affect the flowering process, causing many flowers not to be fertilised and resulting in poor fruit set. Late spring frosts can burn the new season's buds and tendrils. Heavy downpours of rain can cause problems at any point; in spring rain may damage the flowers before they are fertilised; in summer rain can cause botrytis bunch rot when followed by humid days; in autumn, rain can cause ripe grapes to split.

Hail and frost can damage the vines at any time of the season.

Grapes affected by Botrytis cinerea.

Botrytis cinerea

Botrytis cinerea is a fungal disease affecting many plants that grow in wet and humid climates, including grapes. It is known to have a malevolent form where it is referred to as bunch rot or grey rot. In both dry white and red wines, botrytis is considered undesirable. It can change the flavour of the grape and reduce the yields.

The benevolent form is known as noble rot and it is important in the production of sweet wines. Botrytis can make grapes taste sweeter. The fungus pierces tiny holes in the skin of the grape so that juice (water) begins to ooze out and the grape becomes shrivelled and raisin-like. This concentrates the solids left inside the grape such as sugars, fruit acids and minerals. This results in a more intense, concentrated grape that can give a wine aromas of dried apricots and raisins, or marmalade with notes of honey. Botrytised wines are usually sweet and smooth on the palate and can age well.

As New Zealand is a relatively rainy country, botrytis occurs in all wine regions, especially those with a wet and humid weather pattern.

Environmental factors

It is important to match the grapevine to the most advantageous location. For example, Riesling needs less warmth and sunlight than Cabernet Sauvignon in order to fully ripen.

CLIMATE

Historically, vines have shown that they can flourish between the latitudes of 30 to 50 degrees north and south of the equator. This is often referred to as the temperate zone. Altitude of a vineyard site also influences growing conditions and it is possible to grow vines in regions with hot climates and high altitudes, such as Mendoza in Argentina. There are a number of methods used to assess the desirability of a particular microclimate, which is the climate characteristics near or within the vine canopy.

There exist a network of weather stations around the world that record core data such as daily temperatures and rainfall. A degree-day is a method used to classify climatic zones based on the average daily temperatures above 10°C during the vine's growing season. Sunshine hours are a measure of sunshine intensity of a particular area and at the same time are an indicator of cloudiness in an area. The diurnal or daily range measures the variation in temperature between day and night.

The aroma and taste of a wine give a hint as to the climate in which it was grown. Warmer climates often result in aromas of ripe fruit, more body, soft, riper tannins, high alcohol and lower acidity. Wine from cooler climates often has more green fruit, higher acid, and less alcohol, tannin and body.

ASPECT

The direction of a vineyard's slope is known as aspect. This takes into account its orientation to the sun throughout the day. A north-facing aspect is considered an advantage in New Zealand as it influences a vine's ability to ripen grapes. Steep slopes will also affect the drainage of water.

WATER

Water is vital for vines to grow and thrive. Rain and irrigation of water are the two most common ways of providing water to a grapevine. If a vine is exposed to too much water, its grapes can become bloated and their flavour is diluted. Later in the growing season, too much rain can result in grapes splitting, which makes them prone to disease and infection. Irrigation systems are able to manage the reticulation of water around a vineyard.

NUTRIENTS

Nutrients are chemicals required by an organism to live and grow and remain healthy. Nutrients come in many forms, ranging from carbohydrates, fats, proteins and vitamins through to minerals, oxygen and water.

In general, grape vines primarily absorb nutrients directly from the soil using their roots. Some of the most important nutrients, often referred to as minerals, for the health of vines are nitrogen, phosphorus, magnesium, sulphur and potassium. These are dissolved in the soil and taken up by the roots then distributed to the whole plant. Chlorosis is a condition that can occur in iron-deficient limestone-rich soils. It causes the leaves to turn yellow as the affected plant has little or no ability to make carbohydrates through photosynthesis and so may die.

SOIL

Grapevines are drought-tolerant plants that can grow in a wide variety of soils. The composition of the soil influences nutrients available to a vine. For grapevines, it is considered that the poorer the soil, the higher the potential quality of the grape. This is because vines planted in very fertile soils can grow too prolifically, producing masses of canes and leaves at the expense of developing and ripening their fruit. As a consequence soils of low to moderate fertility are preferred.

Loam is soil that is composed of sand, silt and clay. David Jackson[2] uses the following classifications.

Stoneleigh Vineyard, Rapaura.

- **Sandy loams:** Generally these contain a high proportion of relatively large particles (sands). Roots easily penetrate sandy loams because they do not hold a lot of water or nutrients; they become dry in summer, requiring irrigation.

- **Silt loams:** These have smaller particles (silts) that are smaller than sands, hold more water and are usually more fertile.

- **Clays:** These have very fine particles and good water-holding capacity, although some can become very compact and make it difficult for water and roots to penetrate. Clay soils are often difficult to manage and can become soggy in wet weather.

- **Stones and gravels:** When contained in a loam, they reduce the soil's water-holding capacity and the stones make the soil difficult to cultivate.

- **Limestone and chalk soils:** These soils are alkaline in nature. Vineyards may reveal their high lime content by yellowing of grape shoot tips, indicating a deficiency of iron in the soil. They have good water-holding capacity. Champagne, Chablis and Waitaki are good examples.

Pests and diseases

Phylloxera, discussed previously, is a pest that causes disease and deformities. Bunch rot is a fungal disease (See page 60). Other pests include insects known as mealybugs, which like phylloxera are sap-suckers and cause a decline in vine vigour. They also transmit leaf-roll viruses from vine to vine. Insecticide sprays are generally used to control them. Grapevine moths and mites can also damage vines.

BIRDS

Believe it or not, birds (starlings, blackbirds and wax-eyes) are a major pest and can eat an entire grape crop, damaging them in a way that makes the grapes unusable. The most common solution is to cover the vines with a fine netting in late summer. In Marlborough, there is a project to hand-rare the endangered New Zealand falcon or Karearea as it is a natural predator of other pests.

Another tool is air guns that are regulated by timers and shoot into the air at regular intervals during the course of the day.

RABBITS

Rabbits enjoy young vines and nibble on their tender bark; this is why you see new vines enclosed in plastic sleeves.

FUNGI

Botrytis is a grey mould as previously discussed. Powdery mildew (oidium) is a fungal disease that affects all parts of the vine and covers it with a fine dusting of spores. This is usually treated with sulphur-based sprays. Downy mildew (peronospora) develops during warm, humid summers. It requires copper-based sprays to bring under control. Eutypa dieback (dead arm) is a fungal disease that rots the vine wood and is often caused by pruning. This affects the yield of a vine.

VIRUSES

There are a large number of viruses that can infect vines and reduce yield and grape quality. Some viruses enter the vineyard on new plant material supplied by nurseries. The leaf-roll viruses are problematic and can be spread by mealybugs. If badly infected, the vines will have to be destroyed.

Netting of vines to deter birds gorging on the grapes at Mt Difficulty.

Vineyard cycle

SPRING

In spring, the new buds break out from the dormant wood. Flowering will occur about 10–13 weeks after bud break. This is a crucial time as severe frost can be devastating.

1. New plantings at Aurora Vineyard, Bendigo, Central Otago.

2. Bud break.

3. Spring growth on old Chardonnay vines at Neudorf.

4. Tendrils reaching out.

SUMMER

In summer, generally late December to February, the vines are growing quickly and are hungry for water and nutrients. Management of the vine includes shoot thinning and trimming the vine canopy. Leaf plucking encourages growth of fruit rather than foliage and opens up the canopy to limit shading of the bunches. If fruit set has been prolific, a green harvest or bunch thinning may be necessary. This usually happens just before veraison and helps to ensure the vine can fully ripen the remaining bunches. Irrigation is often needed. Care is taken to ensure that disease and pests are not left untreated. Veraison marks the beginning of the ripening process, as sugar levels in the grapes climb and acid levels fall.

1. Prolific growth of foliage and fruit in early summer.

2. Canopy management (leaf plucking) at Herzog.

3. Green harvest at Misha's Vineyard.

4. Deploying netting to prevent birds getting grapes at Herzog.

AUTUMN

In autumn, generally March to May, tasks include tracking and analysing the progress of the ripening fruit and monitoring vines for pests and disease. For most instances, when grapes are deemed to be at their peak ripeness and flavour development, (physiological ripeness) harvest begins. This can be carried out by hand or machine harvester.

WINTER

In winter, generally June to late August, leaves fall and the bare vines become dormant. Vineyard tasks include pruning of the vines, mulching of prunings, repairs of wires and posts, fertiliser application, nursery work, grafting, and preparing spring rootstock.

Pruning

Grapes need to be heavily pruned in order to flourish in the following spring. If they are not pruned each year, a vine can develop too many shoots, which tangle and start to smother the vine. Approximately 90% of the previous season's growth is cut back each winter. Pruning determines the next season's crop, as fruit will develop on new shoots arising from one-year-old canes or spurs.

The method of pruning depends on the variety and its location. In New Zealand, cane pruning it the most common form. Here one, two or four canes are chosen to bear fruit the following spring. Spur pruning is when one or two branches are permanently trained along a trellis and then pruned back to two or three buds.

Picking late harvest fruit.

Pruning, collection of bud wood, Felton Road Vineyard, Central Otago.

Vine trellising systems

Healthy vines just want to grow and ideally need to find something they attach their tendrils to. Training of the vines is how growth is managed. This means the growth can be contained within a structure so that the vine's energy is focused on growing fruit rather than a forest of branches and leaves. The canopy refers to the branches, leaves and fruit of the vine that grows above the trunk.

The density of planting between vines can vary significantly depending on the vineyard. Vines may be planted 900 mm to 2 metres apart. Rows are generally spaced 1.5 to 2.5 metres apart to take into consideration the size of the machinery used in the vineyard for tasks such as spraying, leaf plucking and harvesting. Vineyards with less space between vines are considered better as the vines compete with each other for nutrients and put more energy into growing quality bunches of fruit.

In New Zealand, the most common form of trellising is the Vertical Shoot Positioning (VSP) system, with two or four canes that are replaced each year (see diagram). The Scott Henry system is a variant of VSP and is popular with Sauvignon Blanc growers. It uses the wires to train two canes downward and two canes upward. The result is a veritable hedge that a machine harvester can easily straddle in order to pick the grapes. VSP can also be spur pruned, which means that permanent wood (cordon) and spur provides the new season's fruit.

Bush vines or *gobelet* is the traditional method without any trellising. The vine has a thick trunk and is free-standing. It is low to the ground and the tendrils will often grow over the ground. This system is not generally used in New Zealand.

At Kumeu River, the lyre system is employed where the vines are split in two and trained as two distinct canopies. This helps manage vine vigour.

Pruned vines lie dormant in winter at Stoneleigh Vineyard.

Vertical Shoot Positioning (VSP) system.

Scott Henry trellis system.

Sustainability

In 2007, New Zealand Winegrowers launched an initiative called Sustainable Winegrowing New Zealand (SWNZ).

The goal was to introduce a model of best sustainable practice for both vineyard and winery. They argued that this would also address consumer concerns regarding winemaking and the environment.

Sustainable Winegrowing New Zealand is now an integral part of New Zealand wine production with vineyard participation at almost 100%. Members of SWNZ are committed to a range of environmental, social and employment principles including reducing the use of chemicals, energy, water and packaging, and wherever possible reusing and recycling materials and waste.

Sustainable Winegrowing New Zealand includes:

- A framework for viticultural and winemaking practices that protect the environment while efficiently and economically producing premium wine grapes and wine.
- A programme of continual improvement to ensure companies operate with a goal of enhancing their operational practices.
- A platform for technology transfer so that companies are kept up to date regarding any new technology and its application.
- An external audit structure that has integrity and rigour to comply with market expectations.

See http://www.nzwine.com/sustainability/sustainable-winegrowing-new-zealand/

Seresin Estate, Marlborough.

Yarram Vineyard with companion wild flowers.

Organics and biodynamics

ORGANIC WINEGROWING

Many winegrowers have taken the steps required to become certified as organic and vineyards have recently become one of the fastest-increasing areas of New Zealand's organic production. By 2012, there were over 2500 certified organic vineyard sites and over 100 vineyards growing grapes organically. This equates to 7.6% of all vineyards. Organic Winegrowers New Zealand (OWNZ) has set the goal to have 20% of New Zealand wines certified organic by 2020.

Organic winegrowers strive to cooperate with nature. These wines do not use synthetic chemical fertilisers, pesticides or herbicides but instead rely upon understanding and working with ecological processes and naturally derived products. Organic management aims to build healthy soils by nurturing a diverse, rich soil and insect life and creating a thriving vineyard landscape .

As a result, OWNZ believe that organic and biodynamic growers can produce wines that are a true expression of the land where they are grown.

The International Federation of Organic Agriculture Movements (IFOAM) defines organic agriculture as follows:

Organic agriculture is a production system that sustains the health of soils, ecosystems and people. It relies on ecological processes, biodiversity and cycles adapted to local conditions, rather than the use of inputs with adverse effects.

Organic agriculture combines tradition, innovation and science to benefit the shared environment and promote fair relationships and a good quality of life for all involved.[3]

Millton biodynamic preparations.

BIODYNAMIC WINEGROWING

Austrian philosopher Rudolf Steiner is the father of biodynamics, promoting an environmental and sustainable approach to growing plants and raising animals. His philosophy includes the belief that there exists an energy force related to the cosmic rhythms so that the movements of the moon and planets have a profound influence on the soil, plant and animal life.

While growers often start by gaining organic certification, many move on to biodynamics. James Millton of Millton Vineyards is considered the leader of the biodynamic movement in New Zealand. He was one of the ten original biodynamic winegrowers who founded the French-inspired *Renaissance des Appellations* ('Rebirth of Appellations'). This group adheres to a strict charter of biodynamic practices.

Biodynamics considers the earth and therefore each vineyard as a living organism. Each vineyard activity affects the others. Management is based on the careful observations and the results of tests and analyses.

Vineyards looks to build soil fertility using a range of tools specific to each property. These include:

- Use of biodynamic preparations and sprays to stimulate biological activity in the soil, and improve retention of nutrients.
- Promoting biodiversity through grazing stock and sowing cover crops.
- Stocking with several different animal species to vary grazing patterns and reduce pasture-borne parasites.
- Widening the range of pasture species.
- Planting trees for multiple purposes.
- Crop rotation designed to enhance soil fertility and control weeds and plant pests.
- Recycling of organic wastes where possible, by large-scale composting.
- Changing from chemical pest control to prevention strategies based on good plant and animal nutrition.

CERTIFICATION

All certified organic and biodynamic producers must pass annual audits to ensure compliance with international organic standards. Growers must adhere to organic methods for three years before attaining full certification.

Organisations who certify for organics in New Zealand

- BioGro: http://www.biogro.co.nz
- AsureQuality: http://www.asurequality.com
- Demeter: http://www.biodynamic.org.nz (organics and biodynamics)
- Organic Farm New Zealand: http://www.organicfarm.org.nz

Certified organic wineries

In Marlborough, a group of organic and biodynamic vineyards have joined together to form the Marlborough Natural Winegrowers or Mana Group.

For more information, watch this video: http://naturalwine.org.nz/biodynamics/winter-spring-biodynamics-at-huia-vineyards/

These include:

- Seresin
- Hans Herzog
- Te Whare Ra
- Clos Henri
- Rock Ferry
- Huia
- Fromm

Other organic vineyards

- Black Estate
- Kusuda
- Sato
- Felton Road

Biodynamic vineyards

- Millton Estate
- Rippon
- Pyramid Valley
- Richmond Plains

Information about organics in New Zealand

- Organics Aotearoa New Zealand (OANZ): http://www.oanz.org

Millton vineyard calf.

Compost at Millton.

CarboNZero

Landcare New Zealand launched carboNZero as the world's first internationally accredited greenhouse gas footprint certification under 1SO 14065. In 2006, Grove Mill Winery, owned by the New Zealand Wine Group, became the first certified carbon neutral winery in the world.

CarboNZero accreditation measures the exact carbon emissions of a product or organisation from the beginning to the end of its life cycle. This includes techniques used for soil management, growing vines, production, packaging, shipping and disposal of the empty wine bottles after consumption. It also audits the organisation by assessing the greenhouse gases that it produces, including how staff travel to and from work, national and international air travel and many other issues. The organisation is then able to offset this by the evaluation of procedures used to minimise its carbon footprint, such as the use of renewable energy, recycling, programmes to plant or regenerate forests and wetlands and the purchase of carbon credits.

CarboNZero also aims to be a recognised certification method to verify the integrity of an organisation's claims about greenhouse gas footprint, emissions reduction and carbon neutral claims.

See http:// www.carbonzero.co.nz/

NOTES

1. Quote from Paton, Clive, *The Gumboot Clone*.
2. Jackson, D. (2004).
3. http://www.ifoam.org/en/organic-landmarks/principles-organic-agriculture.

CHAPTER THREE:
White grapes

Sauvignon Blanc, ready to harvest.

WHITE GRAPE DESCRIPTORS

Grape	Primary	Secondary/Tertiary	Palate
Sauvignon Blanc	**Green fruit:** gooseberry, green apple **Herbaceous:** grass, leaves **Citrus:** grapefruit, lemon **Tropical fruit:** passionfruit, pineapple	Grassy, capsicum, asparagus, smoky, flinty	Dry, medium to high acidity, some aged in oak, medium to full-bodied
Chardonnay	**Citrus:** lemon, grapefruit **Stone fruit:** apricot, peach **Tree fruit:** apples **Tropical fruit:** pineapple, rock melon Mineral	**Influence of oak:** vanilla, toast, caramel, toasted nuts **Influence of MLF:** butter, creamy, yoghurt	Dry, medium to high acidity, medium to full-bodied body, creamy mouthfeel
Riesling	**Citrus:** lemon, grapefruit **Tree fruit:** apples, pears **Tropical fruit:** pineapple **If botrytised:** honey, dried apricot and marmalade	With age develops kerosene, petrol notes and toasted nuts, mineral	From dry to very sweet, medium to high acidity depending on style
Gewurztraminer	**Tropical fruit:** especially lychee **Floral:** rose petals, jasmine, geranium Honey Citrus	Spice, musky Mandarin orange	From dry, off-dry, medium low to medium acidity, medium to full-bodied
Pinot Gris	**Stone fruit:** nectarine, apricot **Tree fruit:** pear, ripe apples	Spice, vanilla if oaked Honey	From dry, off-dry, medium, low to medium acidity, medium to full-bodied
Viognier	**Stone fruit:** peaches, apricots **Tree fruit:** pear, apple **Floral:** honeysuckle, hawthorn blossoms	**Stone fruit:** peaches, apricots, citrus, pear Almonds	Dry, medium to full-bodied, medium acidity, oily texture
Chenin Blanc	**Stone fruit:** peaches, apricots **Tree fruit:** pear, apple **Citrus:** lemon **Tropical fruit:** honeydew melon	Lanolin, waxy	Dry to off-dry to medium to very sweet (botrytised) medium acidity, medium to full-bodied
Sémillon	**Herbaceous:** grass, leaves **Citrus:** grapefruit, lemon **Tropical fruit:** passionfruit, pineapple, melon, figs	Lanolin, waxy, hay	Dry to medium to very sweet (botrytised), medium acidity, medium to full body

Sauvignon Blanc

Sauvignon Blanc.

Pronounced: So-veen-yon blonk.

Sauvignon Blanc has become synonymous with New Zealand and the vibrant, herbaceous Marlborough style leads and dominates the New Zealand wine industry.

WHERE IT COMES FROM

It was believed to have come from Bordeaux in France, but now researchers believe it originated from Loire; it was first mentioned in Sancerre in 1783. DNA profiles suggest it is a sibling of Chenin

Blanc and with Cabernet Franc is the parent of Cabernet Sauvignon.[1]

The whites from Bordeaux include both dry Sauvignon Blanc-dominated blends and sweet Sémillon-dominant blends. Sauvignon Blanc is also highly respected in the Loire Valley's dry whites, Sancerre and Pouilly-Fumé, where it is unblended and unoaked.

WHERE IT GROWS

While Sauvignon Blanc comes from France, it is now widely grown around the world. It is grown in Australia, Italy, North and South America and also in South Africa, which is also gaining respect as a high-quality producer of the variety.

New Zealand Sauvignon Blanc is best known for its crisp acidity, intense grassy, herbaceous nose and fruit-driven character.

The grape is early to mid-ripening with vigorous growth, requiring canopy management. It is best not planted in fertile soils. Its small bunches of berries make is prone to botrytis.

COLOUR

Sauvignon Blanc when young is pale lemon with green hints and a watery hue. With age it will develop into a lemon-gold-coloured white wine.

PRIMARY AND SECONDARY AROMAS

The primary fruit aromas of Sauvignon Blanc make it one of the world's most easily recognisable wines. Sauvignon Blanc is associated with pronounced fruit flavours of green apple, passionfruit and pineapple. It can show a herbaceous character of green capsicum and freshly cut grass. These aromas are often associated with Sauvignon Blanc from Marlborough, Chile and South Africa.

The French style of Sauvignon Blanc tends to be less pronounced, with more savoury aromas of green herbs and straw.

When it ages, Sauvignon Blanc can develop aromas of canned asparagus and canned peas.

In cool climates, Sauvignon Blanc displays the aromatic effects of the methoxypyrazines, chemical compounds that produce odours, especially the green herbaceousness. (Methoxypyrazines are present in Cabernet Sauvignon and Cabernet Franc as well.)

Montana's Brancott Estate where the vines were planted east–west rather than north–south meant that one row was more shaded and resulted in high levels of methoxypyrazines.[2]

WINE STYLE

In general, the style is unoaked and fruit driven, although more barrel-fermented and barrel-aged Sauvignon Blanc are becoming available.

Sauvignon Blanc is considered a dry white wine with medium to high acidity. Many consider that Sauvignon Blanc is developed when bottled and is not enhanced by extended bottle ageing. The variety has been described by Jancis Robinson MW as an *exhilarating and easy-to-understand wine'.*[3]

Sauvignon Blanc in New Zealand

High yields and little ageing mean that the grapes are harvested in March–April and able to be released by August–September. This makes Sauvignon Blanc a very profitable wine, especially as it does not require expensive oak barrels for ageing.

VITICULTURE

Sauvignon Blanc produces a different style of wine when grown on stony soils compared to clay soils, both of which Marlborough has in abundance. The vine is vigorous, and the region has high grape yields.

REGIONS

Generally, commercial Sauvignon Blanc is blended from grapes from different vineyards, so it is not easy to determine the different flavour profiles. However, there is a growing trend to make wines from single vineyard sites using oak and malolactic fermentation.

Marlborough

Marlborough remains New Zealand's largest Sauvignon Blanc producer with 90% of all plantings. The 'Marlborough' style is crisp, fresh, aromatic, tangy and flavourful.

TRY THESE WINES

Classic Marlborough Sauvignon Blanc

- Oyster Bay
- Villa Maria Private Bin
- Brancott
- Matua
- Yealands
- Hunters
- Wither Hills
- Saint Clair
- Nautilus
- Giesen

Premium Classic Marlborough Sauvignon Blanc

- Cloudy Bay
- Dog Point Section 94
- Spy Valley single vineyards
- Astrolabe
- Churton
- Vavasour

Barrel-fermented Sauvignon Blanc, indigenous yeast ferment and barrel ageing

- Greywacke Wild Sauvignon Blanc
- Seresin Marama Sauvignon Blanc
- Cloudy Bay Te Koko Sauvignon Blanc
- Mahi Ballot Block Sauvignon Blanc
- Brancott Chosen Rows Sauvignon Blanc

Hawke's Bay

The Hawke's Bay style of Sauvignon Blanc tends to be lighter in aroma, more full-bodied in style and with riper flavours of stone fruit and lower acidity than Sauvignon Blanc from southern regions.

- Te Mata Cape Crest
- Sacred Hill Sauvage
- Quarter Acre
- Unison

Wairarapa

Wairarapa is a similar latitude to Marlborough and makes a similar style.

- Palliser Estate Sauvignon Blanc
- Urlar Sauvignon Blanc
- Martinborough Vineyards Sauvignon Blanc

Nelson

Located north-west of Marlborough, it is influenced by the Marlborough style, although some wines have a more tropical-fruit character.

- Seifried Sauvignon Blanc
- Greenhough Sauvignon Blanc

North Canterbury and Central Otago

Many of their Sauvignon Blanc wines are blended with grapes grown in Marlborough, although a number of vineyards are producing wine off local fruit.

North Canterbury

- Bellbird Springs Sauvignon Blanc
- Fiddler's Green Sauvignon Blanc
- Mt Beautiful Sauvignon Blanc

Central Otago

- Rockburn Sauvignon Blanc
- Misha's Vineyard Sauvignon Blanc
- Amisfield Fume Blanc

Chardonnay

Chardonnay.

Pronounced: Shar-don-nay.

Chardonnay is the classic grape of Burgundy and Champagne. It is widely regarded as responsible for producing some of the world's greatest dry white wines. Chardonnay is versatile because is does not have conspicuous aromas and can take on aromas from the winemaking process.

WHERE IT COMES FROM

Chardonnay is made in a variety of styles. The wine from Chablis, in the north of Burgundy, with its cool climate, is traditionally made in a unoaked style. The grapes are grown on limestone and show high acidity and minerality with fruit flavours of green apples and lemon. The style of Chardonnay from the

Côte de Beaune in the heart of Burgundy reflects the influence of oak in the winemaking, with flavours of butterscotch, vanilla and hazelnut alongside lemon, apples and melon. Here, the best wines are located around a collection of villages and ranked as Grand Cru and Premier Cru. The villages of Puligny-Montrachet and Meursault are considered true expressions of white Burgundy.

Chardonnay is also one of the three key grapes used to make Champagne and high-quality sparkling wines around the world. If often gives the citrus character to Champagne and adds elegance.

WHERE IT GROWS

Today, Chardonnay is grown around the world, primarily because it is easy to grow and make into wine. It is planted extensively in Europe, the United States, Chile, Argentina, South Africa, Australia, New Zealand and Canada.

COLOUR

The colour of Chardonnay ranges from pale lemon green when young to deep gold as an aged wine. The longer Chardonnay spends in the barrel and then in bottle, the greater the colour will intensify from pale to deep, taking on yellow-golden hues.

PRIMARY AND SECONDARY AROMAS

When young, Chardonnay has aromas of stone fruit such as peach, apricot and nectarine. It can also show citrus fruits such as lemons and grapefruit as well as apples. If the wine has been fermented or aged in oak, it can develop notes of butterscotch, caramel, hazelnut, vanilla, spice and cedar.

WINE STYLE

Chardonnay is full-bodied, often reaching medium to relatively high alcohol levels of 13–14%. This high-alcohol content can make Chardonnay seem sweet on the palate even when bone-dry.

Chardonnay can have a neutral aroma and taste when youthful. Chardonnay is generally defined as a dry, full-bodied white wine. Tastes range from subtle to savoury.

Most Chardonnay is made with the intention that it is consumed within the first two to three years, but premium Chardonnay can confidently age beyond 10 years.

Winemaking techniques of lees-stirring, malolactic fermentation, barrel ageing and barrel fermentation influence the style of Chardonnay.

Chardonnay in New Zealand

REGIONS

Chardonnay suits New Zealand's cool climate and is grown in every region with varying degrees of success, depending on the climate and soils in each area.

The most highly regarded Chardonnay comes from Auckland, Hawke's Bay, Nelson, and Marlborough. There are pockets of excellence in Canterbury and Central Otago.

VITICULTURE

Chardonnay is the third most planted grape in New Zealand today. It is interesting to recall that between 1992 and 2003, Chardonnay was the most planted grape in New Zealand. Its influence has been undermined by the fashion of Sauvignon Blanc and more recently, Pinot Gris, however now, Chardonnay is regaining some of that lost ground.

Chardonnay buds early and can be prone to spring frosts. It is easy to grow and ripen and prefers limestone or calcareous (lime or chalky) clay. It may suffer from coulure. It also can develop botrytis and make delicious sweet wines. The Mendoza clone was favoured in New Zealand for its 'hen and chickens' berries with the uneven fruit set reducing yield. This is now being replaced with other Dijon clones.

TRY THESE WINES

Premium barrel-fermented Chardonnay, indigenous yeast ferment and barrel ageing

- Kumeu River Mate's Vineyard
- Mahi Alchemy
- Kumeu River Huntinghill Vineyard
- Te Mata Elston
- Bell Hill
- Felton Road Block 2
- Villa Maria Reserve Barrique fermented
- Ata Rangi Petrie
- Trinity Hill
- Neudorf Moutere
- Black Estate
- Sacred Hill Rifleman's
- Mountford
- Dog Point
- Pyramid Valley Lion's tooth
- Morton Estate Black Label

Other Chardonnay

- Craggy Range Cape Kidnappers
- Hunter's Marlborough
- Waipara Springs
- Martinborough Vineyards
- Escarpment
- Babich Irongate

Unoaked Chardonnay

- Coopers Creek
- TW Estate

Pinot Gris

Pinot Gris.

Pronounced: Pee-no-gree.

Pinot Gris and Pinot Grigio are different names for the same grape. Until 2006, it was known as Tokay Pinot Gris in France, where it is now called Pinot Gris.

WHERE IT COMES FROM

The home of Pinot Gris is regarded as Burgundy, but is also know in Alsace, France as well as southern Germany. It is a mutation of the Pinot Noir grape.

Wines made from the Pinot Gris grape can vary greatly. When made in Italy and Australia they tend to be bone-dry and light compared to those from Alsace, France, which are made in a range of styles from dry to off-dry table wines, as well as sweet dessert wines.

WHERE IT GROWS

Pinot Gris is cultivated widely around the world.

It ripens early but can quickly lose acidity. Thanks to its relatively delicate flavour, often low acidity and potential for high alcohol, Pinot Gris has grown in popularity.

Pinot Gris is one of the four noble grapes in Alsace, France, along with Riesling, Muscat and Gewurztraminer. It is also prolific as Pinot Grigio in northern Italy. It is a popular grape and wine in Oregon in the United States, New Zealand and Australia.

COLOUR

The colour of Pinot Gris can vary widely, which is a hallmark of this grape. As a mutation of Pinot Noir, it is considered to be genetically unstable. For this reason, its skin ranges from white to pink to black and occasionally the wine will be pale lemon and have a gris or grey-pink hue.

PRIMARY AND SECONDARY AROMAS

Pinot Gris displays delicate aromas of stone fruit – pear, nashi pear, ripe nectarine, peach and apricot – as well as some floral notes. Citrus flavours of lemons and limes are also apparent.

Generally, Pinot Grigio and Pinot Gris are not aged for long periods in the bottle. When Pinot Gris is aged, these wines develop aromas of honey, yeast, nuts, lemon and sweet pear.

WINE STYLE

There are good quality wines produced in all regions but Pinot Gris can have varying levels of residual sugar so it is difficult to know how dry the style will be. Pinot Gris is commonly dry, in the more austere style of Pinot Grigio from Italy; off-dry or medium dry, but until you open the bottle, it is difficult to tell.

Pinot Gris has the potential to be relatively full-bodied with medium to low acid and high alcohol. The perfume is fragrant. Some verge on the edge of flabbiness with not enough acidity while others can also be surprisingly high in alcohol.

Pinot Gris in New Zealand

REGIONS

Plantings of Pinot Gris have grown dramatically since 2000. Currently, it is the fourth most planted grape in New Zealand. This rapid upsurge in plantings reflects growth in the country's biggest wine region, Marlborough, where over half of the Pinot Gris is grown. Hawke's Bay is in second place and varying amounts of Pinot Gris are grown in all other wine regions.

VITICULTURE

Pinot Gris is still relatively new to this country and it could be argued that the best clones have yet to be uncovered. Pinot Gris, in many cases, was often top grafted onto less-valued vines. It is an early-budding variety and ripens early in the season. It grows in small bunches with small berries, but ripens easily and therefore has potential for high-alcohol wines. The vine is vigorous, but with variable yields. Pinot Gris can develop botrytis and downy mildew.

TRY THESE WINES

Dry style

These grapes have a naturally higher acidity, which adds a refreshing quality to the wines.

- Bilancia
- Hans Herzog
- Dry River
- Pasquale Wines
- Mahi Ward Farm

Off-dry style

- Te Whare Ra
- Bellbird Spring
- Greystone Waipara Valley
- Brennan
- McDonald Vineyard
- Kumeu River
- Mt Difficulty
- Misha's Vineyard
- Prophet's Rock

Medium-dry style

- Ostler, Waitaki
- The Ned, Marlborough
- Gibbston Valley Wines
- Neudorf Vineyards
- Forrest, Marlborough
- Winegrowers of Ara

Riesling

Riesling.

Pronounced: Rees-ling.

Riesling is considered one of the most ancient German grape varieties and has been used in many crosses, including Muller-Thurgau and Osteiner.

Around the world, Riesling produces consistently fine wines in a variety of styles from bone-dry, highly acidic wines to luscious dessert wines. Many consider Riesling to be underrated and in New Zealand it has its own champions making outstanding examples of this wine.

WHERE IT COMES FROM

Riesling is the principal grape of Germany, but the earliest recorded history of Riesling is in Alsace, France.[4]

WHERE IT GROWS

Today Riesling grows in many countries but it is generally considered that the greatest wine comes from the German regions of Mosel, Rheingau and Rheinhessen. It also flourishes in Alsace, Austria, Australia and New Zealand.

It can tolerate cold, but in order to ripen needs to be planted in good sites.

COLOUR

Riesling tends to be pale lemon with a green hue when young. As it ages, the colour develops from light to deep lemon to gold. This progression of colour can be rapid, especially if the grapes have been affected by botrytis.

PRIMARY AND SECONDARY AROMAS

The aroma can vary widely, depending on the region in which it is grown and the style in which the wine is made. The hallmark aromas of Riesling are lime, lemon and green apples and honeysuckle for dry and medium wines. The sweeter styles of Riesling, often influenced by botrytis, show aromas of honey, lemon, dried apricots, floral aromas and marmalade.

As Riesling ages, it can develop more complex aromas, sometimes with smoke and petrol notes (which can confuse and deter drinkers) with pronounced honey and marmalade characters.

WINE STYLE

Riesling is one of the most versatile grapes because its high acidity balances its sweet aromatic flavours in wines across a spectrum of styles. Riesling can be dry, light to medium-bodied or very sweet. As a consequence, many wine drinkers are often confused about what style of wine to expect from Riesling. To resolve this, many winemakers now state on bottle labels what style of Riesling they are making.

The American based International Riesling Foundation[5] has established guidelines for sweetness that you can now see on some New Zealand bottles. The scale essentially indicates where the wine sits based on the sugar acid balance:

Dry - Medium Dry - Medium Sweet - Sweet

Riesling has the ability to make top-quality wines and the finest examples have the potential to age and improve for several decades in the bottle.

Riesling in New Zealand

In recent years Riesling has been overtaken by Pinot Gris in popularity. Many Riesling vines are being removed from vineyards while others are being top-grafted to become the base of other grape varieties.

Despite this, there an now several Riesling-dedicated events which are being driven by winemakers who are keen to see one of their favourite grapes grow in stature in the public's mind.

REGIONS

Riesling grows in every wine region in New Zealand, but the best-quality versions come from southern areas, including Martinborough, Nelson, Marlborough, North Canterbury and Central Otago. The Auburn vineyard in Central Otago is a Riesling specialist offering six Rieslings that range from a dry style with 5 g/L to sweet with 75 g/L residual sugar.

VITICULTURE

Riesling buds late and therefore is a mid to late-ripening grape. It has a relatively hardy vine structure, which lets it withstand cold temperatures in winter as it prefers a long, slow ripening season to achieve perfect ripeness. This makes it ideally suited to New Zealand's relatively cool southern climates.

TRY THESE WINES

Dry
- Esk Valley Marlborough
- Love Block
- Rippon
- Prophet's Rock

Off-dry
- Dry River
- Greenhough Hope
- Maude Dry
- Mount Edward
- Greenhough
- Palliser Estate
- Nga Waka
- Dancing Water

Medium
- Pegasus Bay
- Forrest Estate
- Sherwood Estate Stratum
- Ataahua
- Martinborough
- Waipara Springs
- Alan McCorkindale, The Mound
- Brancott Waipara

Late Harvest and Noble
- Framingham Noble
- Pegasus Bay Aria
- Mt Difficulty Late Harvest
- Forrest Late Harvest

Gewurztraminer

Gewurztraminer.

Pronounced: Ge-vur-stra-mean-er.

Gewurztraminer is also known as Traminer, with the German word 'Gewurz' (spiced) attached to describe its distinctive fragrant nose.

It is a profusely aromatic wine with the potential for high alcohol, over 14%, and with low acidity.

WHERE IT COMES FROM

DNA profiling shows that Gewurztraminer is not a separate grape variety but a mutation of the grape Savagnin along with Savagnin Blanc (also known as Traminer) and Savagnin Rose, which has pink berries.[6] First records of it date from 1827 in

Rheingau, Germany. It was also thought to be native to the village of Tramin in the Alto Adige region of Italy.

WHERE IT GROWS

Today Gewurztraminer is 20% of the vineyard area in Alsace and one of the four noble grapes along with Riesling, Muscat and Pinot Gris. It is also widely planted in Baden and Pfalz in south-western Germany. It grows in many wine-growing parts of the world and is a success especially in areas of New Zealand, the United States and the cooler regions of Australia.

COLOUR

The skin of the Gewurztraminer berry is light pink, but in the glass the wine is generally lemon in colour, although sometimes with a hint of pink.

PRIMARY AND SECONDARY AROMAS

Lychees, floral, rose petals, jasmine, Turkish Delight are terms used to describe Gewurztraminer, which is one of the most instantly recognisable wines. Gewurztraminer is pungent and aromatic and this distinctive character makes it easy to remember and recognise.

WINE STYLE

Gewurztraminer can be dry, off-dry, medium to very sweet. It often has an oily texture with medium to low acidity and is sometimes described as 'flabby'. Alcohol can often reach 13–14% and Gewurztraminer wines are considered full-bodied. Early-picked grapes can result in a more neutral wine.

Gewurztraminer in New Zealand

Gewurztraminer has a low profile in New Zealand. Many consider that its name, which is difficult to pronounce, is partly responsible for this. In the last decade, plantings have declined as other grapes have risen in popularity. New Zealand Gewurztraminer consistently achieves high levels of success in international wine competitions.

REGIONS

Gisborne, Hawke's Bay and Marlborough are the most successful regions for Gewurztraminer in New Zealand. These regions are diverse and do not share similarities in soil or climate, which leads to the conclusion that Gewurztraminer has the potential to make high-quality wines in most regions.

VITICULTURE

Gewurztraminer is sensitive to wind, which can destroy the vine's ability to flower in spring. It can show vigorous growth, but crops vary enormously from one season to the next, especially if the spring has been windy. It buds early and therefore is at risk from spring frosts.

TRY THESE WINES

- Dry River
- Blackenbrook
- Waimea
- Johanneshof Cellars
- Te Whare Ra
- Lawson's Dry Hills
- Ellero
- Brancott Marlborough
- Rockburn Wines
- Stonecroft
- Cloudy Bay
- Zephyr Vineyard
- Vinoptima

Viognier

Viognier.

Pronounced: Vee-on-e-ay.

Viognier is growing in popularity around the world as an alternative full-bodied wine to rival Chardonnay. Since this time, it has flourished, not only in France but around the world. Viognier greatly benefited from the collective admiration of a group of Californian winemakers and enthusiasts, known

as the Rhone Rangers. Since the 1980s, they have become the champions of the 22 Rhône Valley grape varieties and have lifted the profile of all these wines.

WHERE IT COMES FROM

Viognier is from France's northern Rhône Valley and the important appellation of Condrieu, where it has been grown for centuries. Due to its low productivity, by 1968 Viognier plantings had declined to just 14 hectares.[7]

DNA profiling indicates that Viognier is either a half sibling or grandparent of Syrah. Even today, in Côte Rôtie it is common for a small amount of Viognier to be fermented with the Syrah.

WHERE IT GROWS

From its traditional home in Condrieu, Viognier is widely planted in the Rhône Valley and Languedoc-Roussillon regions of France and is important in the blend of white wines permitted for the Côte du Rhône and Côte du Rhône Villages. It is extensively planted in California and Australia where over 500 vineyards make Viognier. It is slowly growing in New Zealand.

COLOUR

The colour of Viognier ranges from pale lemon when young to pale gold as it ages.

PRIMARY AND SECONDARY AROMAS

Honeysuckle, gingerbread, apples, pears, apricots, peaches, nectarines and delicate white flowers aromas are all part of Viognier's aromatics, and they are usually strong in young wines. As it ages, Viognier tends to lose its intense fruity and floral aromas as the aromas develop away from these primary characteristics.

WINE STYLE

Viognier generally makes a dry wine, of medium to high acidity with an oily mouthfeel. Like Chardonnay, it is considered full-bodied but offers a more aromatic nose. It is often high in alcohol, around 14%, with relatively low acidity. Generally, it is best consumed when it is young.

Viognier in New Zealand

Viognier plantings are growing in New Zealand, from just 15 hectares in 2002 to around 140 hectares in 2013.

REGIONS

While Viognier is still a relatively minor grape variety in this country, its profile with winemakers and wine drinkers is lifting. Most Viognier in New Zealand is made with grapes grown in Gisborne, Hawke's Bay, and Marlborough.

VITICULTURE

Like the Gewurztraminer grape, Viognier is relatively sensitive to wind, making it a poor cropping grape in windy seasons. For this reason, it is relatively unreliable commercially in areas which are wind-prone.

It is an early-budding vine and so can be prone to spring frosts. It ripens well and can therefore result in wines of lower acidity and high alcohol.

TRY THESE WINES

- Hans Herzog
- Te Mata Zara
- Elephant Hill
- Craggy Range
- Millton
- Waimea
- Mudbrick Reserve
- Alpha Domus Wingwalker
- Villa Maria Omahu Gravels
- Babich

NOTES:

1. Robinson, J. (2012), p. 953.
2. Courtney, C. (2003), p. 132.
3. Robinson, J. (2013), p. 360.
4. Robinson, J. (2012) p. 888.
5. For more information, please look at: http://drinkriesling.com/tastescale
6. Ibid., p. 960.
7. Ibid., p.1144.

CHAPTER FOUR:
Red grapes

Pinot Noir.

RED GRAPE DESCRIPTORS

Red Grapes	Primary	Secondary/Tertiary	Palate
Pinot Noir	**Red fruit:** cherries, strawberries, raspberries **Black fruit:** black cherries, plums	**Earthy:** mushroom **Oak influence:** spice, vanilla, caramel toffee	Dry, medium to high acidity, medium-soft tannins, medium-bodied
Cabernet Sauvignon	**Black fruit:** blackcurrant, black cherry **Herbaceous:** green capsicum, mint, eucalyptus	**Oak influence:** spice, vanilla, caramel toffee, oak, tobacco	Dry, medium to high acidity, if cool climate high tannins – will soften with age, medium to full-bodied
Merlot	**Black fruit:** black plums, blackberry, black cherry **Red fruit:** red cherries	**Oak influence:** spice, vanilla, caramel toffee, oak, tobacco, chocolate	Dry, medium acidity, medium to full-bodied, medium tannins
Syrah/Shiraz	**Black fruit:** blackberry, black cherry, ripe plum **Herbal:** menthol, lavender, mint, white pepper	**Oak influence:** black pepper, black liquorice, violet, chocolate, vanilla	Dry, low to medium acid, medium ripe tannins, medium to full-bodied
Cabernet Franc	**Red fruit:** crushed berries, plums, red cherry **Herbaceous:** herbal, green capsicum	**Oak influence:** tobacco, vanilla and spice	Dry, medium acid, medium-bodies, medium to high tannins
Malbec	**Black fruit:** blackberry, blackcurrant, ripe plum, **Herbaceous:** herbal, green capsicum	**Oak influence:** black pepper, black liquorice, vanilla	Dry, medium to high acid, medium to high tannins, medium to full-bodied

Pinot Noir

Pinot Noir.

Pronounced: pee–no–nwar.

Pinot Noir is the noble red grape of Burgundy's Côte d'Or. It is the progeny, along with 156 other grape varieties, of Pinot, considered one of the oldest grape varieties in the world, thought to have been in existence for around 2000 years.

Pinot Noir produces wine that is softer on the palate without the harsh tannins of young Bordeaux blends, and this makes it very appealing.

Pinot Noir, like Chardonnay, plays an important role in the production of Champagne and for many sparkling wines around the world.

WHERE IT COMES FROM

Pinot is the parent of Pinot Noir, Pinot Meunier, Pinot Blanc and Pinot Gris. Once thought to be separate varieties, DNA profiling shows that these Pinots have the same genetic fingerprint. In other words, rather than refer to them as from the 'pinot family' they are in fact mutations within a single variety.[1]

The first record of Pinot Noir in Burgundy dates from 1375. Planted extensively in the small villages of the Côte de Nuits, the Pinot Noir grape is known for its elegance and finesse. Since the 1930s, many of these villages have attached the name of the most famous wine in their appellation to the village name. For example, the village of Gevrey is known as Gevrey-Chambertin and Puligny is known as Puligny-Montrachet. In Burgundy, the most celebrated wines are produced from vineyard sites ranked as Grand Cru or Premier Cru.

WHERE IT GROWS

Today Pinot Noir has spread all over the world, with varying degrees of success. It tends to produce the best grapes when grown in relatively moderate to cool climates. It is an early-ripening grape and a cool climate enables it to ripen without losing aroma or acidity. The best-quality Pinot Noirs are often associated with limestone soils. Beyond Burgundy, it is grown in other areas of France and to increasing acclaim in Oregon, USA, as Spatburgunder in Germany, Eastern Europe, many areas of New Zealand and the cooler regions of Australia.

COLOUR

The colour of Pinot Noir wine can vary considerably. It ranges from light to deep ruby, depending on the quality of the fruit. If the wine is a simple, light Pinot Noir for everyday drinking, it will tend to be relatively light in colour whereas top-quality Burgundies can be deep ruby.

PRIMARY AND SECONDARY AROMAS

Pinot Noir offers red-fruit aromas such as strawberries, raspberries, red cherries, red plums and red currants; there are also black-fruit aromas from riper climates, such as blackberries, cherries and beetroot, coupled with dried herbs, spice, mushrooms, and other savoury nuances.

As the wines ages, it develops aromas of 'forest floor', hints of truffles, mushrooms, savoury and complex earthiness.

WINE STYLE

Pinot Noir is made in two broad styles.

The lighter style is characterised by ripe red summer fruits and has medium body, with tannins often described as ripe or silky. It may have medium to high acids. The best wines can be deceptively light in colour but have intense, multi-layered flavours and great length.

A more full-bodied style, from riper fruits, will often have more black fruit character and higher alcohol.

As a young wine, Pinot Noir can develop quickly but depending on its style it can age unpredictably.

Pinot Noir in New Zealand

Pinot Noir is the most planted red grape and the second most planted grape throughout New Zealand. In 1990 it was grown in a only a few vineyards, but today it is an enormously successful wine, both in terms of domestic sales and exports all over the world.

It is estimated that 10% of the crop is usually destined for sparkling wine production.

REGIONS

Wairarapa, Marlborough, Nelson, North Canterbury and Central Otago are the most important regions for Pinot Noir, in terms of plantings of vines, and the quality and quantity of wine made from them.

MARTINBOROUGH

Martinborough was the first New Zealand region to gain international recognition for the quality of its Pinot Noir. The original vineyards were planted during the 1980s, along a seam of limestone, near the town of Martinborough. Dry River, the Martinborough Vineyard and Ata Rangi were some the area's pioneers. Their goal was to make Pinot Noir of grand cru quality and, to this end, they imitated the vineyard management and winemaking practice of Burgundy.

These wines are often considered to show more red cherry, strawberry and raspberry – summer fruit characters balanced with subtle flavours associated with ageing in French oak.

Impressed with the success of Martinborough Pinot Noir, other regions were quick to follow.

CENTRAL OTAGO

In the mid 1990s, planting in Central Otago took off. As New Zealand's premier tourism destination, the growth in boutique vineyards added a new experience for international tourists visiting Queenstown.

The fruit-driven style of Central Otago Pinot Noir is often associated with ripe black cherries, dried herbs and plums, higher alcohol and sweet spice from ageing in new oak barrels. It is this 'friendly' style that has made the wine so successful and highly sought after. Wooing Tree Blondie is a *Blanc de Noir*, a white wine made from 100% Pinot Noir.

MARLBOROUGH

Pinot Noir in Marlborough was originally planted to make sparkling wine, as an important component in Montana's hugely successful Lindauer blend.

Over time, Pinot Noir from Marlborough has grown in reputation, with a variety of styles being produced, some of a lighter style while others show dense plums and beetroot flavours and aromas.

NORTH CANTERBURY

There are also outstanding examples of Pinot Noir grown in North Canterbury. Pegasus Bay remains a leader and innovator with a number of styles of Pinot Noir produced from its winery. Pyramid Valley and Bell Hill, located near Waikari along with a number of vineyards in the Waipara Valley prove that this region can produce elegant Pinot Noir.

VITICULTURE

Pinot Noir has relatively thin skin and is considered a cool-climate grape as it ripens too quickly in hot climates. It is an early-budding variety and is at risk from spring frosts. Pinot Noir is early-ripening, produces small bunches and may suffer from coulure.

The Pinot Noir vine is prone to mildew, rot and viruses including fanleaf and the leaf-roll virus.

If over-cropped, Pinot Noir immediately appears as a dilute, thin, acidic wine, a pale reflection of its true potential. Pinot Noir has also spawned a vast number of different clones with different growth characteristics.

TRY THESE WINES

Central Otago

- Quartz Reef
- Peregrine
- Valli
- Mt Difficulty
- Burn Cottage
- Mount Edward
- Gibbston
- Akarua
- Rockburn
- Rippon
- Forrest Otago
- Misha
- Grasshopper Rock Earnscleugh

Waitaki, North Otago

- Valli
- Waitaki Braids
- Ostler
- Pasquale

Marlborough

- Clos Henri
- Fromm Clayvin
- Isabel Estate
- Seresin
- Churton
- Dog Point
- Auntsfield
- Nautilus
- Marisco

North Canterbury

- Pegasus Bay
- Waipara Springs
- Fancrest
- Mountford
- Greystone
- Black Estate

Martinborough

- Ata Rangi
- Martinborough
- Escarpment
- Schubert
- Urlar
- Julicher
- Te Kairanga

Nelson

- Neudorf
- Richmond Plains
- Greenhough

Super-premium

- Peregrine, the Pinnacle
- Felton Road Block 3
- Gibbston Valley Reserve
- Sato
- Kusuda
- Pyramid Valley Cowley
- Wooing Tree Sandstorm
- Bell Hill

Cabernet Sauvignon

Cabernet Sauvignon.

Pronounced: Caberr-nay So-veen-yon.

Cabernet Sauvignon is considered the world's most renowned grape for the production of fine, long-lived red wine. DNA tests confirm that it is the progeny of Cabernet Franc and Sauvignon Blanc and a sibling of Merlot.

It is blended with different combinations of Merlot, Cabernet Franc, Malbec and Petit Verdot. This is famously known as the 'Bordeaux blend' or in Britain as 'Claret'.

WHERE IT COMES FROM

Bordeaux is the home of Cabernet Sauvignon, but it was only at the end of the eighteenth century that Cabernet Sauvignon began to make a significant impact in the vineyards of Bordeaux in the Médoc and Graves.

WHERE IT GROWS

Cabernet Sauvignon grows almost everywhere wine is made in the world today. It is highly transportable and flourishes in the hot climates of California, Chile, Argentina and Australia as well as moderate Mediterranean zones such as Tuscany and Languedoc-Roussillon. In New Zealand, it is at its best in Hawke's Bay.

COLOUR

In a glass, Cabernet Sauvignon is dark ruby red, verging on fuchsia to deep purple when young. As it ages, Cabernet Sauvignon based reds acquire a bricky hue that moves towards garnet as the wine ages.

PRIMARY AND SECONDARY AROMAS

Cabernet Sauvignon has highly recognisable aromas of black fruits, including blackcurrants, plums, black berries and black cherries. It also has a herbaceous quality (methoxypyrazine) of green capsicum, leaves and fresh herbs. The primary aromas influenced by oak develop into a complex range of cedar, spice and dried herbs with notes of earth and black olives in aged Cabernet Sauvignon.

WINE STYLE

Cabernet Sauvignon styles are generally full-bodied and full-flavoured with medium to high tannins and acidity. In top wines, it has the ability to age and improve for decades, moving from primary bright fruit flavours when young through to savoury, multi-layered ones when aged.

One of the keys to Cabernet Sauvignon's ability to age is its small, thick-skinned berries, which have a relatively high ratio of pip to pulp. This accounts for the wine's strong tannic structure, which gives it the potential to age.

Cabernet Sauvignon in New Zealand

Cabernet Sauvignon has declined in New Zealand since 2003, as Merlot has become the preferred grape. Merlot ripens earlier than Cabernet Sauvignon and is currently considered the more reliable red Bordeaux grape in this country.

Cabernet Sauvignon is the seventh most widely planted grape variety in New Zealand. Historically, Cabernet Sauvignon was planted everywhere from Northland to Central Otago and while small numbers of vines are still grown in some regions, it is generally accepted that Cabernet Sauvignon is more suited to the hotter, drier Hawke's Bay climate, compared with the relatively cool climate and high rainfall in other areas.

Today, it is difficult to find a straight Cabernet Sauvignon in New Zealand although Church Road's McDonald Series is making an excellent example.

REGIONS

Hawke's Bay is the region in which most New Zealand Cabernet Sauvignon is grown, thanks to its moderate climate.

Waiheke Island also has a reputation for high quality Cabernet Sauvignon, although numbers of grapes planted are relatively small compared to Hawke's Bay. The variety thrives in the Gimblett Gravels.

VITICULTURE

Cabernet Sauvignon has small bunches and grapes with such thick skins that when growing they can appear blue rather than red or purple.

It is vigorous and thrives on well-drained gravel soil It can suffer from fungal diseases, including eutypa dieback and powdery mildew. Cabernet Sauvignon is late ripening and if not fully ripe can show harsh unripe tannins and pronounced herbaceousness.

BORDEAUX BLENDS

Cabernet Sauvignon, Merlot, Cabernet Franc, Malbec and a little known grape, Petit Verdot, are often blended together in different ways and with different proportions of grapes. These are commonly referred to as Bordeaux blends. From the 1960s, Cabernet Sauvignon was the dominant force however, by the late 1990s, Merlot had become the foremost grape.

For the winemaker, a Bordeaux blend requires mastery of the blending process; balancing the components of the different wines, blended together to ensure an outstanding outcome.

TRY THESE BORDEAUX BLENDS

- Te Mata Coleraine
- Craggy Range Sophia
- Vidal
- Pask
- Stonecroft
- Stonyridge Larose
- Te Awa
- Trinity Hill
- Unison
- Clearview Enigma
- Mission
- Morton Estate
- Mills Reef
- Babich, the Patriach
- Villa Maria offers many Bordeaux blends

Merlot

Merlot.

Pronounced: Mer-low.

Merlot is best known as the partner of Cabernet Sauvignon and its genetic sibling. When blended together, they produce the most famous wines of Bordeaux. Merlot offers ripe fruit to balance Cabernet Sauvignon's herbal notes, tannins and acidity.

WHERE IT COMES FROM

Merlot has the longest recorded history in Bordeaux, France, where it is the key red grape in St Emilion and Pomerol. Its full body, medium tannins and rich fruit flavours are prized for adding flesh to the more structured bones of Cabernet Sauvignon from Bordeaux's left-bank region.

WHERE IT GROWS

Merlot grows easily in a wide range of climates, soils and aspects; for this reason it is planted almost everywhere wine is made today.

The true expression of Merlot wines are made in the wineries of St Emilion and Pomerol in Bordeaux. Merlot is now extensively planted in Italy and an important component of Super Tuscan wines. Other important Merlot areas include California, Australia, Chile, South Africa and New Zealand.

COLOUR

Merlot is medium ruby to deep purple in hot climates. As the wine ages it develops brick tones.

PRIMARY AND SECONDARY AROMAS

As a young grape, ripe red and black plum aromas characterise Merlot when it is young and the use of oak can add an overlay of cedar, spice and 'cigar box' type aromas. As it ages, Merlot is best known for adding softness to blends, both in terms of taste and aroma, but as a single varietal wine, Merlot can develop aromas of mocha, earth and spice as it ages.

WINE STYLE

The style of Merlot is influenced by whether it is used as a blending grape or as a single varietal wine. Generally, it is considered to produce full-bodied, high-alcohol wines.

As a blending grape, Merlot adds flesh, body and softness to the bones of the more structural Cabernet Sauvignon and Cabernet Franc.

As a single varietal wine, Merlot is full-bodied with medium to high tannins, soft red to black fruity flavours and, in a good vintage, it can have great length and ability to age in the bottle.

Merlot in New Zealand

Merlot has grown in importance, reflected by the increasing number of vines planted in New Zealand over the past decade. It is the fifth most planted grape in New Zealand. Its growth is largely due to its more reliable viticultural habits as it ripens earlier than Cabernet Sauvignon.

REGIONS

Hawke's Bay is home to most of New Zealand's Merlot grapes and only small amounts are planted in other regions including Waiheke Island.

VITICULTURE

Merlot ripens earlier than Cabernet Sauvignon and also generally yields larger quantities of grapes per vine. The vine suits clay and limestone soils and shows moderate to vigorous growth; it is early budding and prone to coulure and at risk from spring frosts and drought

TRY THESE WINES

- Church Road McDonald Series
- Matua Single Vineyard Matheson
- Pask
- Alluviale
- Crossroads Milestone
- Trinity Hill
- Villa Maria Reserve Gimblett Gravels Merlot
- Esk Valley Winemakers Reserve Merlot Malbec Cabernet Sauvignon
- Alpha Domus
- Te Awa

Syrah/Shiraz

Syrah or Shiraz.

Pronounced: Sea-rah/She-raz.

Syrah and Shiraz are the same grape. The French Syrah is the great red wine of Côte Rôtie, Hermitage and Cornas in the Northern Rhône Valley. Shiraz is Australia's greatest contribution to the world of red wine. There is however, an enormous difference in style between these wines. The DNA profile is not conclusive about it origins but it could well appear that Pinot is its great grandparent.[2]

WHERE IT GROWS

Syrah's home is the Rhône Valley where it is often blended with Grenache and Mourvedre. It is widely planted in the Languedoc- Roussillon area with nearly 70,000 hectares in France. Syrah, with its ability to age and improve for decades in the bottle, has spread to many countries including Italy and Spain.

In Australia, Shiraz is the bigger, bolder, fruitier expression the grape. Shiraz is Australia's most planted grape with over 40,000 hectares. It is successfully grown in California, Argentina, Chile and South Africa while New Zealand is a relatively new player.

COLOUR

Deep ruby to fuchsia colour when young, it holds this colour for several years before softening into lighter shades of ruby as it ages.

PRIMARY AND SECONDARY AROMAS

Syrah and Shiraz have youthful aromas of dark black fruit, including black cherries, black plums, black berries, violets, chocolate and black and white pepper. As the wine ages, it may begin to show leather, cedar, spice and vanilla components, reflecting the influence of oak. When grown in hot climates such as Australia it can show herbal mint and eucalyptus characters and the flavours of sweet cooked jam.

The tannins in Syrah, are softer than in Cabernet Sauvignon. As it ages, Syrah and Shiraz develop aromas of violets, coffee and spice.

WINE STYLE

Syrah and Shiraz are considered medium to full-bodied wines with medium to high tannins and high alcohol. The best wines can age for decades, softening over time from tannic to silky reds.

Syrah/Shiraz in New Zealand

Syrah first appeared as a separate grape variety in New Zealand wine records in 1999 with just 40 hectares. Alan Limmer experimented with Syrah at Stonecroft, located in the Gimblett Gravels, and encouraged other winemakers to follow. Today, plantings of Syrah account for over 400 hectares of vines as its reputation for producing high-quality has grown.

REGIONS

Hawke's Bay and Waiheke Island are the two regions best known for Syrah in New Zealand. Relatively small amounts grow in tiny pockets of other regions, most notably in the Wairarapa and also in Marlborough.

VITICULTURE

Syrah/Shiraz needs a warm climate to ripen properly and develop its most desired flavours. The Gimblett Gravels of Hawke's Bay and Waiheke Island have proved the most reliable climates for ripening Syrah in New Zealand.

Syrah is a vigorous grower and needs careful trellising to avoid chlorosis and botrytis bunch rot. It is unsuited to fertile soils with high lime content. Syrah/Shiraz is a mid-ripening variety and berries are small and tend to shrivel when ripe.

TRY THESE WINES

Trinity Hill's Homage is made from Syrah cuttings from Jaboulet's La Chapelle vineyard in Hermitage and Viognier from Les Jumelles in Côte Rôtie. John Hancock was invited to work under the legendary winemaker, Gerard Jaboulet in 1996. The first vintage of Homage was made in 2002. Along with Bilancia La Collina and Craggy Range Le Sol, it is considered one of New Zealand's great wines.

- Bilancia La Collina
- Craggy Range Le Sol
- Passage Rock Reserve Waiheke Island
- Vidal Legacy Series Gimblett Gravels
- Elephant Hill
- Te Mata Bullnose
- Stonecroft (these vines originated from a single clone rescued from Te Kauwhata by Alan Limmer)
- Giesen *The Brothers* Marlborough Syrah
- Sacred Hill Halo
- Fromm Syrah
- Mills Reef
- Mission

Malbec

Malbec.

Pronounced: Mal-beck.

Known originally as Côt in the region of Cahors in south-west France, Malbec is also known for playing a minor role in Bordeaux blends. It is referred to as 'the black wine' in Cahors. Today it is most famous as Argentina's great red.

WHERE IT COMES FROM

Malbec is from the department of Lot in south-west France. Its DNA profile show it to be a half-sibling of Merlot.[3]

In recent years, plantings have declined in France, but in other areas of the world the grape has flourished.

WHERE IT GROWS

Argentina boast over 26,000 hectares of Malbec and it commonly appears as a single variety. In Argentina, Malbec makes outstanding red wines in their relatively hot climate and is responsible for the growth in quality export of wine from that country.

Chile, Cahors and Bordeaux remain key regions but it is often blended. It is also grown in California, South Australia and New Zealand.

COLOUR

Deep purple to black; this colour is one of the wine's most recognisable features.

PRIMARY AND SECONDARY AROMAS

Dense black fruit such as plums, tones of dried dusty, leaves, green pepper and savoury, earthy aromas characterise Malbec when youthful, developing into earthier characteristics as the wine's aroma evolves. As it ages, the influence of oak can introduce pronounced spice, vanilla and caramel characteristics.

WINE STYLE

Full-bodied and dry, Malbec is characterised strongly by intense tannins, good acidity and a black spice aroma and flavour.

Malbec in New Zealand

Malbec is growing in importance in New Zealand and is now often noted on the label of many wines. Plantings have continued to grow from a tiny base of just 25 hectares nationally in 1998 to 140 hectares in 2013, as winemakers discover the strengths of this intensely tannic grape.

REGIONS

Hawke's Bay is home to nearly all of the Malbec planted in New Zealand.

VITICULTURE

Malbec is sensitive to frost, downy mildew, rot and phomopsis blight (a fungal disease). A mid-ripening grape, it is prone to coulure.

TRY THESE WINES

- Pask
- Villa Maria
- Brookfields
- Cooper's Creek
- Mills Reef
- Mission Estate
- TW Dr Rod

Cabernet Franc

Cabernet Franc.

Pronounced: Caberr-nay fronk.

Cabernet Franc buds and ripens earlier than Cabernet Sauvignon and is an important structural component of red Bordeaux, especially in St Émilion such as Château Cheval Blanc.

WHERE IT COMES FROM

DNA profiling shows that Cabernet Franc and Sauvignon Blanc are the parents of Cabernet Sauvignon. Cabernet Franc is now believed to have originated in the Basque region of northern Spain. Cabernet Franc is also a parent of Merlot and the grape Carménère.

WHERE IT GROWS

Cabernet Franc is an important red blend used in the blend from Bordeaux. It is also the most important grape grown in the Loire Valley. It is found around the world in regions that make wines inspired by the famous châteaux of Bordeaux.

COLOUR

Its colour can range from ruby to purple. More often it is used as a blending grape.

PRIMARY AND SECONDARY AROMAS

The aromas of Cabernet Franc are primarily of red fruit and cherries, and with a mild herbaceousness it can tend to be slightly green. It is 'paler, lighter, crisper, softer and more obviously aromatic' than Cabernet Sauvignon.[4] It can be vegetative-smelling due to its methoxypyrazines.

WINE STYLE

Cabernet Franc is medium-bodied but often with medium to high acidity and some bright red-berry fruit and herbaceous characters. It has moderate tannins, making it easier drinking and ideal for red-wine blends. Cabernet Franc is also used to make rosé in the Loire Valley, where it produces a relatively light salmon-pink wine.

Cabernet Franc in New Zealand

Cabernet Franc is a relatively minor grape in New Zealand, with plantings having decreased over the past decade. It is most commonly used, as in Bordeaux, as a blending grape rather than to make single varietal wines.

REGIONS

Most Cabernet Franc in New Zealand is grown in Hawke's Bay followed by Auckland. Both regions are respected for their high-quality Bordeaux blends. Elsewhere, there are pockets of small plantings compared to other red grapes.

VITICULTURE

Cabernet Franc is a vigorous-growing vine that suits clay and limestone. Ripening mid-season, its small berries are similar to Cabernet Sauvignon but it does bud and mature earlier. The small berries are prone to coulure, but ripen easily.

TRY THESE WINES

- Clearview Estate (made from 26-year-old, hand-harvested vines but only in the best years)
- Crossroads Vineyard
- Pyramid Valley
- Bridge Pa
- Sileni Estates

Alternative grape varieties

Viticulturists continue to experiment with lesser known grape varieties to see how they respond to New Zealand conditions. The master of alternative grapes must be Hans Herzog with over 26 different varieties grown on his 11 hectare Marlborough estate. Experimentation has driven him to top graft vines that he is not satisfied with and replace with cuttings from other vines (see p. 56). Coopers Creek is also an innovator when it comes to trialling alternative grape varieties.

ALTERNATIVE WHITE GRAPES

White wines flourish in New Zealand's cool climate and provide the greatest opportunity for new varietals.

These include Marsanne, Roussanne, Petit Manseng from France; Verdelho, one of the traditional grapes of Madeira from the Portuguese island off North Africa.

Gruner Veltliner

Gruner Veltliner is the famous grape of Austria. It is dry with lemon citrus and neutral minerality. Examples include: Waimea Gruner Veltliner, The Doctors, Yealands, Babich, Quartz Reef.

Arneis

Whether we know it or not, Arneis now accounts for over 30 hectares of vines. Arneis is from Piedmont in Northern Italy where is unoaked with neutral flavours of citrus and pears. Examples include: Coopers Creek, Herzog, Trinity Hills Gimblett Gravels Arneis, Matawhero, The Doctors, Mount Edward.

Albarino

The classic grape from Galicia, on the western coast of Spain and northern Portugal is rapidly growing in New Zealand. It is fermented dry with lemon grass and honey flavours. Examples include: Matawhero, Coopers Creek SV Bell-Ringer Gisborne Albarino offers lots of potential.

Pinot Meunier.

Try these wines:

- Rippon Osteiner
- Esk Valley Verdelho
- Churton Petit Manseng
- Coopers Creek Gisborne Marsanne Allison
- Waimea Sauvignon Gris
- Brancott Sauvignon Gris
- The Doctors' Petit Manseng
- Dancing Water Scheurebe

OLD FAVOURITES

Some grape varieties have gone out of fashion, but their star may rise again. Muller-Thurgau and Reichensteiner from Germany are in this category. These grapes include:

Sémillon

The French grape Sémillon, from Bordeaux, is also grown throughout New Zealand and while plantings are small this grape is widely known. It is often blended with Sauvignon Blanc or made into noble wines:

Examples include: Pegasus Bay Sauvignon Sémillon, Clearview Sémillon, Alpha Domus.

Muscat varieties

Muscat grapes come in many forms and can be green or red in colour. There is a Muscat family with Muscat Blanc à Petits Grains considered the parent of Muscat of Alexandria and grandparent of the fashionable Italian grape, Grillo. Historically the Muscat varieties were very important in New Zealand but today are mostly lost in blends or as a component of sparkling wine.

Examples include: Brackenbrook Nelson Muscat, from grapes grown in their Moutere Hills estate with12 g/L residual sugar. Millton Muskats@dawn, Villa Maria S Series Moscato.

Chenin Blanc

The Millton Te Arai Chenin Blanc from Gisborne is probably the most consistently famous New Zealand Chenin Blanc and has a loyal following.

Other Chenin Blancs include: Mt Difficulty Long Gully Chenin Blanc with 50 g/L residual sugar. Esk Valley Moteo is a dry style from Dartmoor Valley and they also make a dessert wine with 143 g/L residual sugar.

Pinot Blanc

Pinot Blanc is a noble grape of Alsace and is generally a medium-bodied wine with moderate acidity. Its delicate aromas and more neutral personality make it an excellent wine to go with many foods.

Examples include: Escarpment Pinot Blanc from Te Muna, Pyramid Valley Kerner Pinot Blanc from Marlborough, Mount Edward from Central Otago. Greenhough Hope Pinot Blanc is made from vines planted in the 1970s.

ALTERNATIVE RED GRAPES

New Zealand is predominantly a white-wine producing country, but that has not deterred many winemakers in both the North and South islands from experimenting with unusual grape varieties which come from various places, from France, Italy and Spain to Russia.

These include: Barbera, Lagrein, Sangiovese, Marzemino, Montepulciano and Nebbiolo from Italy; Garnacha and Tempranillo from Spain; Gamay and Tannat from France; St Laurent and Zweigelt from Austria; Pinotage from South Africa and Saperavi from Russia.

Try these wines:
- Hans Herzog Montepulciano
- Stonecroft Zinfindel
- Trinity Hill Gimblett Gravels Tempranillo
- The Doctors' St Laurent
- Stonyridge Grenache
- Villa Maria Grenache
- De la terre Tannat
- TW Carménère
- Hans Herzog Zweigelt
- Trinity Hill Touriga Nacional

Pinotage

Pinotage, from South Africa, can be smooth and berryish and considered an answer to Beaujolais, or be more seriously oaked. It was widely planted in Gisborne during the 1970s but is now hard to find.

Try these wines:
- Babich Pinotage
- Muddy Water Pinotage

NOTES:

1. Robinson, J. (2012), p. 805–6.
2. Robinson, J. (2012) p. 1023.
3. Ibid., p. 272.
4. Ibid., p. 155.

CHAPTER FIVE:

Making wine

1. *Machine harvesting, Brancott Estate.*

2. *Harvesting at Villa Maria in Marlborough.*

3. *Hand harvesting at Greywacke.*

4. *Harvesting Sauvignon Blanc at the Yarrum vineyard, Marlborough.*

1. *Press at Dog Point, Marlborough.*

2. *Sorting Pinot Noir at Dog Point.*

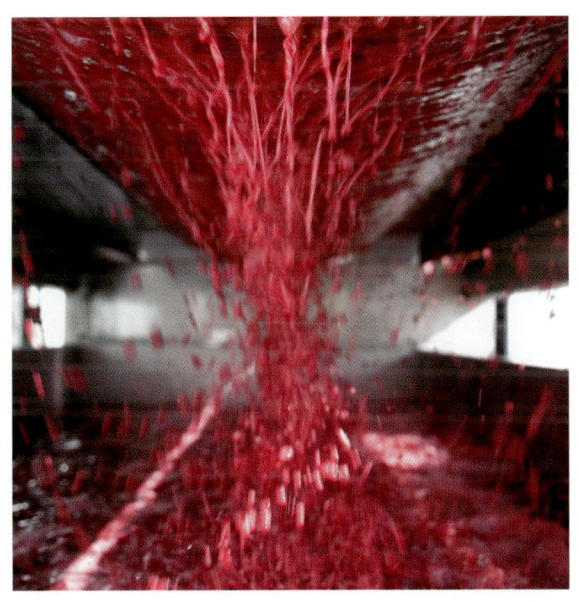

3. *Pressing.*

4. *Barrel maturation.*

Winemaking terms

ALCOHOLIC FERMENTATION

Fermentation is the conversion of sugar into ethanol (ethyl alcohol) by the action of yeast, a process that also produces heat.

Sugar + Yeast = Alcohol + Carbon Dioxide (CO₂) + heat

During this process, the grapes' sugar is converted to alcohol and carbon dioxide gas is given off.

RIPENESS

Sugar content is a decisive factor in wine quality. Ripeness is measured by various means (including sugar, acid and pH) to determine when to pick the grapes. One measure of ripeness is a scale of sugar content referred to in the United States and New Zealand as Brix, in France as Baumé and in Germany as Oechsle.

Physiological ripeness

The 'physiological' ripeness in grapes is a term used to describe a broad range of factors that determine ripeness, including ripeness of tannins and the development of other phenolic compounds that contribute to the colour, flavour and aroma of wine.

Crushing: This breaks up the berries to free the sugars in the juice for fermentation. It must be done carefully so that the pips are not broken, which can result in bitterness. Traditionally, crushing was done by foot.

Destemming: This can take place via the machine harvester as it 'picks' the grapes or with a crusher-destemmer in the winery.

Pressing: This separates the skin and pips from the juice.

Ripe grapes.

Whole-bunch pressing: Grapes are pressed while still attached to their stems.

Whole-bunch inclusion: Whole bunches or clusters (or a proportion of them) are included in the fermentation. The cap is more aerated with stems and helps let some heat escape.

Chaptalisation: This involves the addition of sugar or concentrated grape must (crushed grapes) or rectified grape must (a concentrated solution of natural grape sugars) to the grape juice before or during the fermentation process to increase the final alcohol level of a wine. This is more common in cool climates.[2]

Acidity: Achieving a balanced acidity is an important consideration in the winemaking process. For under-ripe grapes it is possible to chaptalise or de-acidify by using a variety of techniques. It is also common to add acid during the winemaking process to correct the balance in very ripe grapes.

Indigenous or wild yeasts: These are yeasts that are found naturally on the skin of grapes.

Barrel fermentation: A process in which fermentation takes place in a barrel.

Plunging the cap: During fermentation of red wine, this process involves manually pressing down the cap of grapes that has formed on top of the fermenting wine. For many, this has been replaced with **pumping over**, a process where the fermenting wine is drawn off the bottom of the tank and pumped up and on to

the cap that has formed on top. This helps break up the cap or alternatively, the fermenting wine can be left to 'percolate' through the cap.

Racking: This is the process of moving grape must or wine from one tank or barrel to another, leaving the residue behind.

Malolactic fermentation (MFL): Bacterial fermentation that converts malic acid to lactic acids and has the effect of softening and reducing acidity and results in diacetyl (butterscotch, butter aroma). All red wines undergo MFL but only some whites, most notably Chardonnay where it produces a buttery, hazelnut character.

Lees-stirring: The sediment that is left after fermentation is known as the lees or yeast sediment. Some wines, particularly Chardonnay, are left to rest on the lees with regular stirring of the sediment. This gives the wine a fuller body with flavours of bread or biscuits and more complexity.

Blending: This refers to blending wines to produce a wine with different characters, e.g. blending Merlot with Cabernet Sauvignon to produce a softer wine. Winemakers also blend different components of wine such as those with free-run juice with juice from pressed grapes.

Clarification: This filtering process makes wine clear and bright by removing suspended solids such as proteins and yeasts.

Fining: In this process, the wine is cleared by adding agents such as egg whites, bentonite clay, gelatine or casein. This causes the particles of yeast, protein or other matter to adhere to the fining agents and sink to the bottom of the tank or barrel.

Tartrate stabilisation: This involves chilling wine to around −2°C prior to bottling in order for the potassium bitartrate (cream of tartar) to crystallise and precipitate (fall out) and thereby prevent tartrate deposits forming in the bottle after it is sealed.

Filtration: In most instances, the wine is filtered to remove suspended particles before bottling. This may be done using a variety of filtration methods. A wine that is labeled 'unfiltered' will not have gone through this process.

Oak maturation or oak ageing: Wine is left in oak barrels to mature, soften and develop. Oak can also impart woody flavours, tannins and richer textures. Premium oak barrels come from forests in France (e.g. Nevers, Limousin and Tronçais). Other barrels may be made of American, Russian or Slovenian oak.

Bottle ageing: The wine is left to age and develop in the bottle without air, e.g. vintage port, Champagne, classed wines from Bordeaux.

Maturation: A young red wine can taste raw, harsh and acidic. Leaving it to mature allows the tannins to soften in red wine. Freshly made white wines can taste out of balance and additional maturation can allow them to develop.

Malolactic fermentation in spring.

White wine process

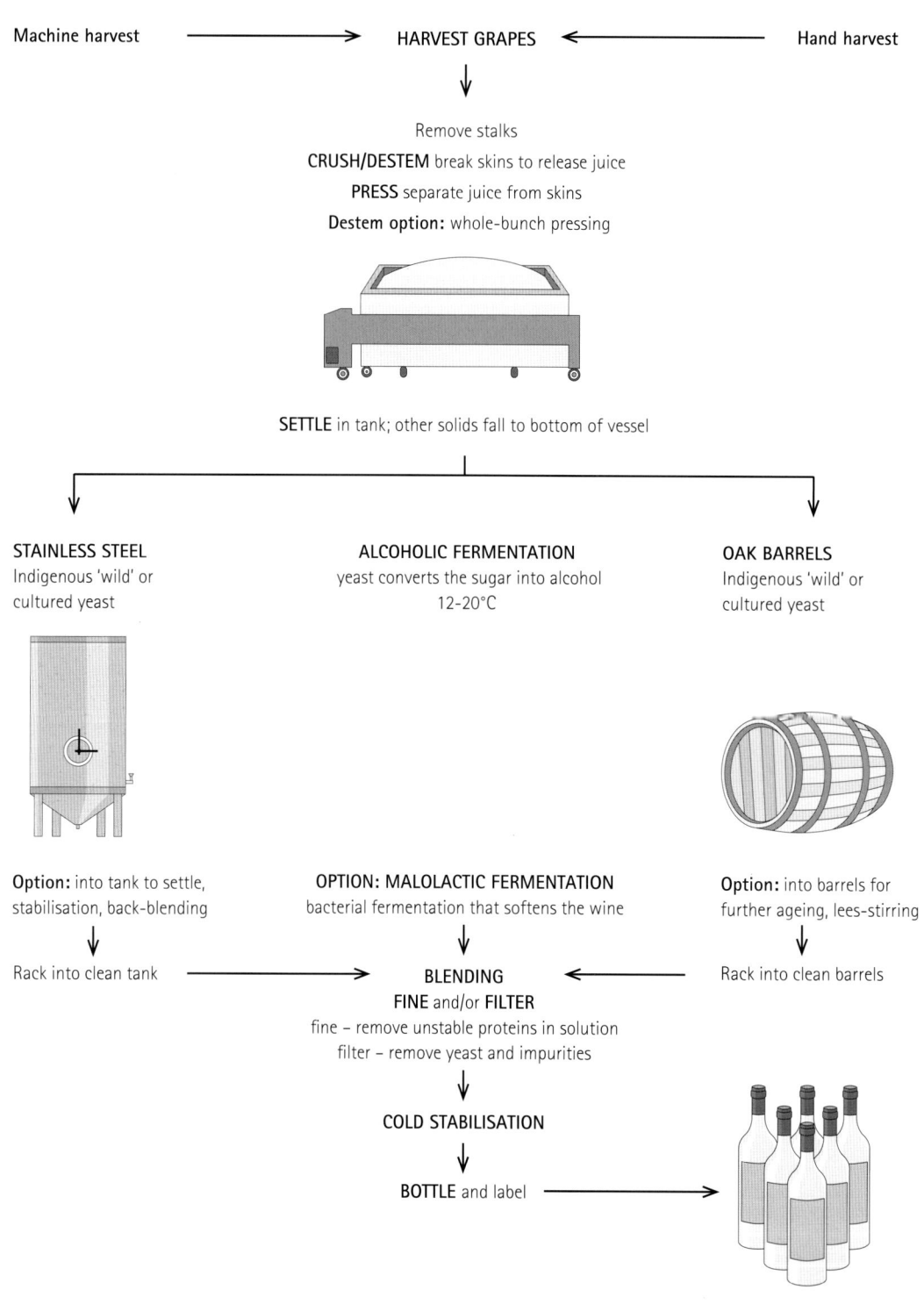

Machine harvest → HARVEST GRAPES ← Hand harvest

Remove stalks
CRUSH/DESTEM break skins to release juice
PRESS separate juice from skins
Destem option: whole-bunch pressing

SETTLE in tank; other solids fall to bottom of vessel

STAINLESS STEEL
Indigenous 'wild' or
cultured yeast

ALCOHOLIC FERMENTATION
yeast converts the sugar into alcohol
12-20°C

OAK BARRELS
Indigenous 'wild' or
cultured yeast

Option: into tank to settle,
stabilisation, back-blending

Rack into clean tank →

OPTION: MALOLACTIC FERMENTATION
bacterial fermentation that softens the wine

Option: into barrels for
further ageing, lees-stirring

Rack into clean barrels

BLENDING
FINE and/or **FILTER**
fine – remove unstable proteins in solution
filter – remove yeast and impurities

COLD STABILISATION

BOTTLE and label →

RESIDUAL SUGAR

The sweetness in wine is known as residual sugar. Unlike grape juice, which is measured in Brix, the sugar in the finished wine is measured in grams per litre. As well as sugar (carbohydrate), 70–85% of a grape is water. The most important acids include malic, tartaric and citric acids.

Residual sugar is retained in a wine to ensure a better balance between the natural sweetness and acidity. Riesling, made in a cool region from grapes with high acidity, may contain well over 30 grams per litre of residual sugar, but will feel dry on the palate.

Stop fermentation

This occurs when the fermentation is halted to preserve some natural sugar in the wine. This is done by the addition of sulphur or by chilling down the wine to kill off the yeast. With Port, alcohol is added to the fermenting wine to kill the yeast and stop the fermentation. Stop fermentation is used for wines that are medium or off-dry.

Susse reserve (back blending)

With this process sterilised, unfermented grape juice is added to the wine after fermentation. This increases the natural sweetness of the wine, dilutes the alcohol and can introduce more fruit aromas.

Red wine process

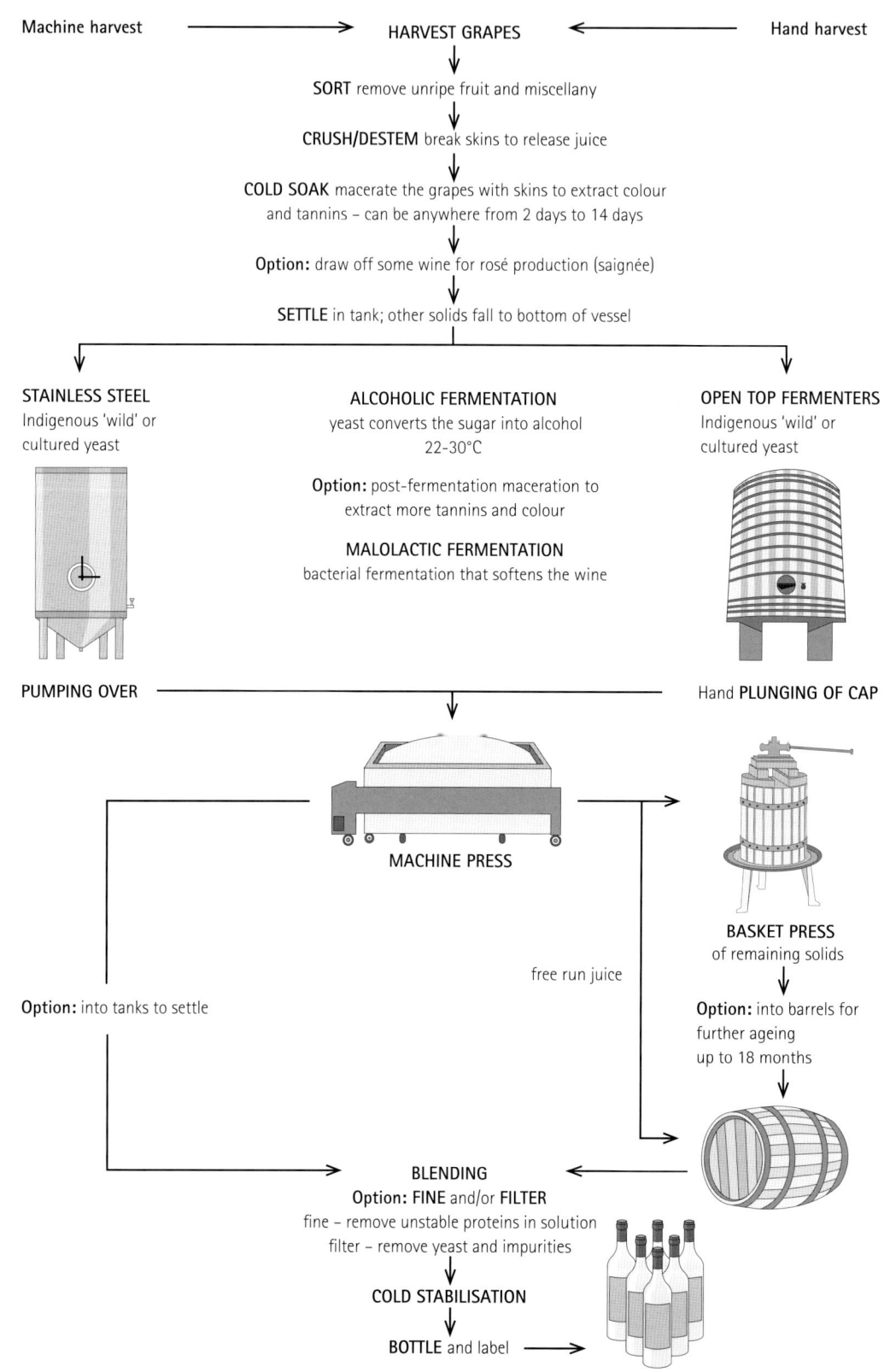

Machine harvest → **HARVEST GRAPES** ← Hand harvest

SORT remove unripe fruit and miscellany

CRUSH/DESTEM break skins to release juice

COLD SOAK macerate the grapes with skins to extract colour and tannins – can be anywhere from 2 days to 14 days

Option: draw off some wine for rosé production (saignée)

SETTLE in tank; other solids fall to bottom of vessel

STAINLESS STEEL
Indigenous 'wild' or cultured yeast

ALCOHOLIC FERMENTATION
yeast converts the sugar into alcohol
22-30°C

Option: post-fermentation maceration to extract more tannins and colour

MALOLACTIC FERMENTATION
bacterial fermentation that softens the wine

OPEN TOP FERMENTERS
Indigenous 'wild' or cultured yeast

PUMPING OVER —————————— Hand **PLUNGING OF CAP**

MACHINE PRESS

Option: into tanks to settle

free run juice

BASKET PRESS
of remaining solids

Option: into barrels for further ageing up to 18 months

BLENDING
Option: FINE and/or **FILTER**
fine – remove unstable proteins in solution
filter – remove yeast and impurities

COLD STABILISATION

BOTTLE and label →

CARBONIC MACERATION

Carbonic maceration is a method of anaerobic (without oxygen) fermentation for red winemaking associated with the region of Beaujolais in France. It involves placing uncrushed grapes in a sealed tank and injecting a layer of carbon dioxide over the grapes that forms a barrier between the grapes and the air. In the absence of oxygen, the berries begin an intracellular fermentation in which some alcohol is produced along with other compounds.

The grapes at the bottom of the tank are crushed by gravity and can undergo conventional fermentation. Yeast is then introduced to finish off the fermentation. This process results in wines that taste fresh and are dominated by primary fruit with low tannin levels.

There are many variations of this process.

It is common for winemakers to use a mixture of whole bunches and de-stemmed grapes when making wine such as Pinot Noir. The whole bunches often sit at the bottom of the vat and remain intact under-going their own carbonic maceration. Meanwhile the crushed berries ferment as normal.

TRY THESE WINES
* Te Mata Gamay Noir
* Rippon Gamay Noir

Rosé

Rosés can be made in a number of ways. The most common method in New Zealand is to draw off juice after 12–24 hours of maceration, when its colour starts to deepen, as part of the red wine process. In France, this is known as saignée or bleeding of the juice. The wine is then fermented as a white wine.

Have you ever wondered why we have such delicious rosés coming from our Pinot Noir producers?

TRY THESE WINES

Craggy Range Gimblett Gravels Rosé is a blend of Merlot and Syrah grapes. It is whole bunch pressed and fermented with wild yeasts in old French oak barriques.

* Felton Road Vin Gris (Pinot Noir)
* Rockburn Stolen Kiss (Pinot Noir)
* Passage Rock Rosé (Syrah, Merlot, Malbec)
* Woollaston Pinot Rosé
* Richmond Plains Monarch Rosé (Pinot Noir)
* Amisfield Rosé (Pinot Noir)
* Locharburn Pinot Rosé
* Bridge Pa Drama Queen Rosé (Syrah)
* Cambridge Road Arohanui (Pinot Noir, Syrah)
* Esk Valley Rosé (Merlot, Malbec)

Sweetness in wine

Late harvest grapes ready for harvesting.

Wines can be made sweeter in a number of ways. Sometimes this is necessary to bring a wine into balance particularly if it has high levels of acidity. In white wines from cool climates, we often find a small amount of residual sugar is retrained and these wines are often referred to as off-dry which falls somewhere between dry and medium dry. The table below gives an overview.[1]

The simplest method is to stop the fermentation early before the yeast has consumed all the sugar in the grape juice. Another option is to add in a sweet component (*Susse Reserve* or back blending) of unfermented grape juice which will also introduce more of the fresh primary aromas.

Dessert wines

Dessert wines are the ultimate expression of ripe grapes. Grapes are left to hang on the vine, until they become over-ripe and shrivelled, perhaps covered in mould, and appear almost past using. The season and weather conditions of a particular year will determine the levels of ripeness in the grapes and the styles of dessert wine that can be made. These take two main forms:

Late harvest

Grapes that are left on the vine to ripen and start to shrivel and resemble raisins are known as late-harvest grapes. The berries are often picked individually.

Noble rot

When conditions are right, with damp misty mornings followed by warm, dry afternoons, some grapes will develop *Botrytis cinerea* or noble rot. The fungus attacks the skin and the grapes begin to shrivel and their juice is drawn off. This has the effect of concentrating the sugar in the berry. These nobly rotten grapes look completely unappetising. It is hard to imagine that they are capable of producing such rich and delicious wine with flavours of dried apricot, ripe pineapple and orange marmalade.

Grapes most susceptible to *Botrytis cinerea* include Riesling, Chenin Blanc and Sémillon.

As botrytis will often affect only part of a bunch of grapes, it is common for the harvest to involve the selection and picking of individual berries in a number of passes through the vineyard at different times.

RESIDUAL SUGAR OVERVIEW

Residual sugar (grams/litre)	Classification	Varieties
Less than 4 g/L	Dry	Chardonnay, Sauvignon Blanc and all reds
About 5-12 g/L	Off-dry	Riesling, Gewurztraminer, Pinot Gris
About 13 -45 g/L	Medium or medium sweet	Riesling, Pinot Gris, most white cask wines
45 and over g/L	Sweet	Late-harvest and dessert wines

Sweet wines, made with such grapes, are highly revered and include some of the most famous and long-living wines in the world: *Tokaji* from Hungary, *Sauternes* from Bordeaux, *Selection de Grains Noble* and *Vendange Tardive* in Alsace; *Beerenauslese* and *Trockenbeerenauslese* in Germany and Austria.

Icewine

Icewine, known in German as *Eiswein*, refers to sweet wine made from grapes that have been frozen on the vine. The grapes are pressed while frozen, and the ice (water content in the juice) is removed, leaving concentrated grape must. This method relies on a heavy frost occurring during the final stage of ripening. Each season is different and it is not predictable when this will happen, if at all. Generally, icewine does not have the influence of botrytis.

In some years, the conditions prove suitable to make icewine in Central Otago.

Cryoextraction

This process imitates the natural conditions in order to make icewine. Grapes are frozen so that when pressed the water content will be removed, leaving the least frozen berries with the highest sugar-rich juice. There are variations on this process.

Examples of these styles include Akarua Alchemy which is made from Riesling and Gewurztraminer and has 218 g/L residual sugar. Te Mania Koha is made from Riesling and Lincoln Ice Wine is made from Gewurztraminer.

Passito

Passito is an Italian term for grapes dried in the sun after the harvest and prior to fermentation. Traditionally, the grapes are dried on straw mats. This process enables the sugar to become more concentrated. Pasquale Passito is made from dried Pinot Gris and Riesling grapes.

Port method

In the Port method, alcohol is added during the fermentation to kill off the yeast and retain a level of sweetness. This is called fortification. The French wine style, Vins doux naturels, often made with the Muscat grape is made in a similar way.

New Zealand's most widely recognised fortified wine is Mazuran Director's Port NV which contains wines from 1975. Torlesse Reserve Port from Waipara is from a solera started in 1992. Clearview Sea Red and Trinity Hill Touriga are Port style wines. Millton Mistelle Fortified July Muscat is a New Zealand style of vins doux naturels.

TRY THESE WINES

In New Zealand, there are many fine dessert wines made in a variety of styles.

Late harvest

- Clearview Late Harvest Chardonnay
- Alluviale Tardif Late Autumn Harvest Hawke's Bay Sauvignon Blanc
- Villa Maria Cellar Selection Marlborough Late Harvest Sémillon
- Seifried Sweet Agnes Riesling
- Cloudy Bay Late Harvest Riesling
- Osawa Late Harvest Gewurztraminer

Noble Rot or Botrytis

- Johner Noble Pinot Noir
- Pegasus Bay Finale Noble Sémillon
- Pasquale Shrivel Riesling
- Villa Maria Reserve Marlborough Noble Riesling
- Framingham Noble Riesling
- Pegasus Bay Encore Noble Riesling
- The Doctors Noble Chenin
- Vinoptima Noble Gewurztraminer

Closures

The closure that is used to seal a wine bottle has an important influence on how a wine develops.

Cork

Cork is the traditional method of closure. Cork is thick, pliable and lightweight. It provides a seal that allows tiny amounts of oxygen to slowly enter the wine over time and promote bottle ageing. Cork is also biodegradable. In recent years, cork closures have become controversial because of the incidence of cork taint and concerns of premature oxidation of a wine. As a consequence, many new closure options have become available to winemakers and research and debate continues as to the best methods of sealing a bottle of wine.

Cork taint

Cork taint, also known as 2,4,6-trichloroanisole (TCA), is transferred via the cork to the wine. Cork taint is often characterised as a dirty, dusty, mouldy, wet cardboard, or wet dog smell and the wine will be described as 'corked' or 'corky'. It is believed to have its origins in the chlorine used to wash the cork bark as part of the production process but may also come from tainted barrels in the winery.

Innovation in cork manufacturing techniques has led to agglomerate corks, made from the ground up crumbs of cork, then moulded into a cork shape. The French company Diam offers a 'technical cork' by boiling, grinding and cleaning the cork and then shaping it into the correct shape. This is growing in popularity in New Zealand.

Screwcaps

The search to find a reliable alternative to cork closures has involved much trial and experimentation. Screwcaps provide an inert closure that creates a seal from the air and in doing so, preserves the fresh fruit character of a wine. Leading brands in include Stelvin and Guala.

Screwcap Initiative

Poor quality corks, imported into New Zealand and Australia in the late 1990s, resulted in a high incidence of cork taint. As a consequence, a group of high profile winemakers decided to collectively change to screwcap closures. The Screwcap Initiative was introduced to promote widespread education of the benefits of screwcaps and gain consumer acceptance of the change.

http://www.screwcapinitiative.com

Synthetic corks

These are plastic corks and come in many forms. Usually wines using these closures should be consumed within one year of bottling. Sometimes these wines throw off reductive odours such as rubber, boiled cabbage or tinned vegetables.

Sparkling wine process

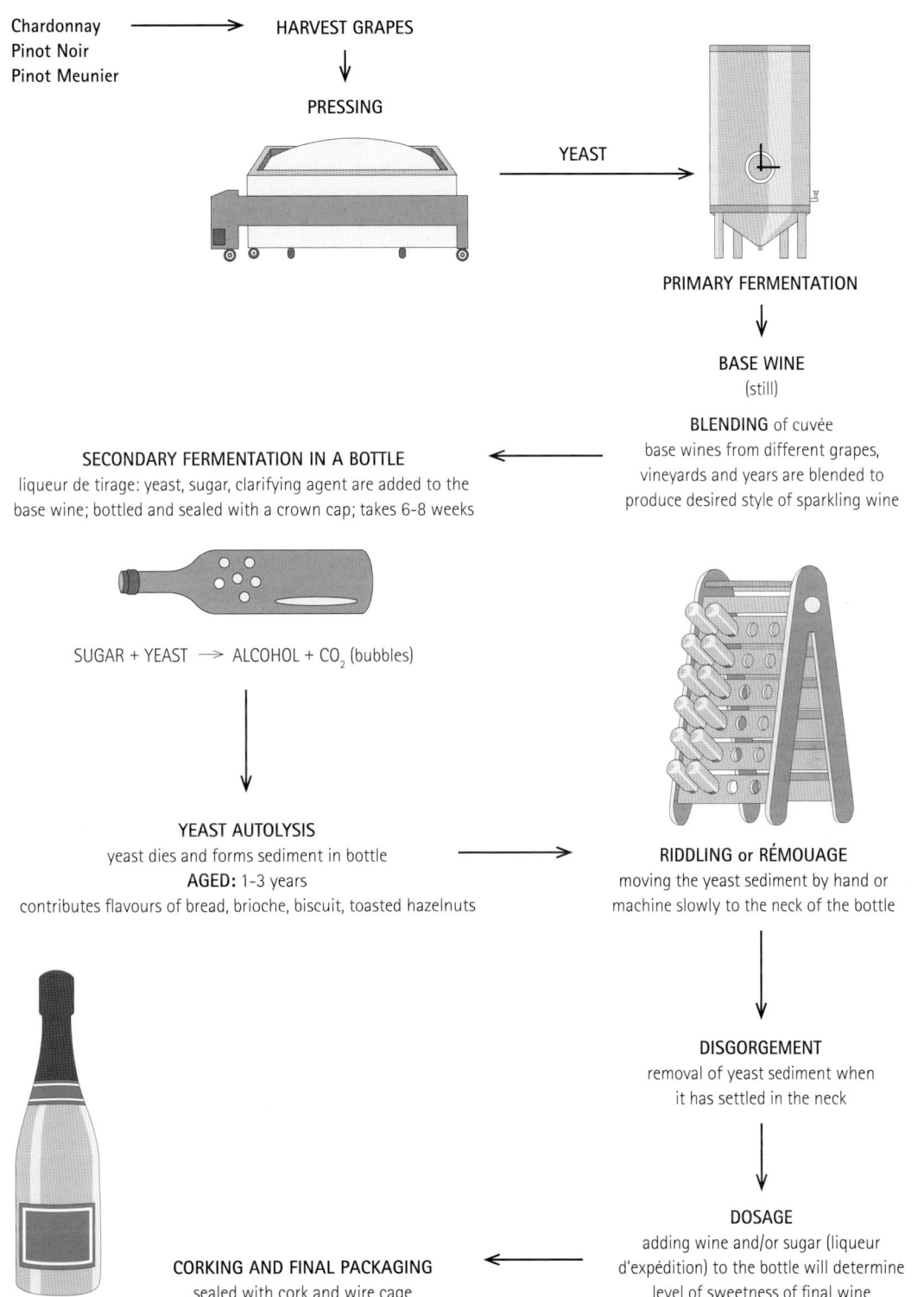

Chardonnay
Pinot Noir
Pinot Meunier

HARVEST GRAPES

PRESSING

YEAST

PRIMARY FERMENTATION

BASE WINE
(still)

BLENDING of cuvée
base wines from different grapes,
vineyards and years are blended to
produce desired style of sparkling wine

SECONDARY FERMENTATION IN A BOTTLE
liqueur de tirage: yeast, sugar, clarifying agent are added to the
base wine; bottled and sealed with a crown cap; takes 6-8 weeks

SUGAR + YEAST $\rightarrow$ ALCOHOL + CO_2 (bubbles)

YEAST AUTOLYSIS
yeast dies and forms sediment in bottle
AGED: 1-3 years
contributes flavours of bread, brioche, biscuit, toasted hazelnuts

RIDDLING or RÉMOUAGE
moving the yeast sediment by hand or
machine slowly to the neck of the bottle

DISORGEMENT
removal of yeast sediment when
it has settled in the neck

DOSAGE
adding wine and/or sugar (liqueur
d'expédition) to the bottle will determine
level of sweetness of final wine

CORKING AND FINAL PACKAGING
sealed with cork and wire cage

Sparkling wine

METHODE TRADITIONELLE OR TRADITIONAL METHOD

This is the method used in the Champagne region of France where the second fermentation takes place in the bottle; the same bottle that is eventually bought by the consumer.

Chardonnay, Pinot Noir and Pinot Meunier are the grapes of Champagne and are traditionally made into a dry wine. A special blend of wines is then made to form the cuvée which will undergo a second fermentation in the bottle. Included in the wine is the liqueur de tirage, a mixture of sugar, yeast and clarifying agent. It is closed with a crown cap. The second fermentation in bottle takes 6–8 weeks and the carbon dioxide generated by the yeast dissolves into the wine, creating the sparkle. The yeast dies and forms a sediment in the bottle called lees. The wine is then aged for 9 months to 3 years (or more) where it undergoes autolysis; the exposure to the lees in the wine. Autolysis contributes flavours such as bread, brioche, biscuit and toasted hazelnuts to the wine and is the defining characteristic of quality wines made using this process.

The final steps involve rémouage or riddling in which the bottle is turned so that the lees moves slowly to the neck of the bottle. This can be done by hand with bottles place horizontally in a pupitres and twisted twice each day for a 6–8 week period. Mechanised gyropalette systems are common to perform this task.

The bottle is then disgorged (dégorgement) and the sediment removed from the neck of the bottle. This is done by submerging the neck in brine and freezing the portion to be removed. The crown cap is removed and the frozen lees ejected. The next step is dosage; the addition of wine and/or sugar (liqueur de expedition) to replace the ejected portion. The level of sugar that is added will determine the level of sweetness of the final wine (see box).

Bottles of Cloudy Bay, Pelorus.

STYLES OF METHODE TRADITIONELLE

- Non-vintage (NV): Wine blended from more than one vintage.
- Vintage: Wine from a single year.
- Rosé: Blend of red and white wine; usually pale pink or salmon colour.
- Blanc de blanc: 'white of white'; wine made from 100% Chardonnay.
- Blanc de noir: 'white of black'; wine made from Pinot Noir and/or Pinot Meunier.

Prestige Cuvée: usually a vintage wine, perceived as the best cuvée of the Champagne house and the most expensive.

LEVELS OF SWEETNESS

- Brut nature or zero dosage: Dry, less than 3 grams per litre of residual sugar, with no sugar added at dosage.
- Extra brut: Bone-dry, 0–6 grams per litre of residual sugar.
- Brut: Dry, less than 12 grams per litre of residual sugar.
- Extra-sec: Dry to medium, less than 12–17 grams per litre of residual sugar.
- Sec: Medium dry, 17–32 grams per litre of residual sugar.
- Demi-sec: Medium sweet, 32–50 grams per litre of residual sugar.
- Doux: Sweet, from 50 grams per litre of residual sugar and upwards.

OTHER SPARKLING WINE PROCESSES

Transfer method

Wine is fermented in the bottle then disgorged into a tank, filtered and rebottled. The removes the costly process of riddling and disgorgement. The process is often referred to as bottle-fermented.

Tank or Charmat method

Dry wine is placed in a sealed tank with sugar, yeast nutrients, and a clarifying agent. The second fermentation takes place, and the wine is then filtered and bottled under pressure. Wines made this way lack the yeasty character imparted by autolysis. Examples include Sparkling Muscat, Riesling, and Prosecco.

Asti method

In this method there is no base wine. Grape must is fermented in a pressurised tank and the carbon dioxide is allowed to escape until a level of 6% alcohol is achieved. Fermentation continues, with the capture of carbon dioxide bubbles, and is stopped when alcohol level reaches around 7%.

Carbonation

Carbon dioxide is injected into a still wine to create bubbles. This method is the cheapest option and is not considered desirable for quality wines.

TRADITIONAL METHOD: TRY THESE WINES

Vintage:

- Quartz Reef Vintage 2009: 87% Chardonnay, 13% Pinot Noir, aged four years on its lees.
- Johner 2009 Chardonnay, aged on lees for 24 months.
- Deutz Blanc de Blanc 2008 Marlborough Cuvée

Non-vintage:

- Cloudy Bay Pelorus NV.
- No1 Family Estate Cuvée Number 8: 65% Pinot Noir, 35% Chardonnay.
- No1 Family Estate Cuvée Number 1: 100% Chardonnay (Blanc de Blanc).
- Johanneshof Blanc de Noirs Brut: 100% Pinot Noir.
- Woollaston Mahana NV Brut but with 2.7 g/L residual sugar would be considered Brut nature, 50% Chardonnay, 50% Pinot Noir.
- Nautilus Marlborough Cuvée NV: 70% Pinot Noir, 30% Chardonnay, aged on less for 36 months.
- Akarua Brut NV (Bannockburn): 71% Pinot Noir, 29% Chardonnay.
- Osawa Prestige Collection NV, 54% Pinot Noir, 46% Chardonnay.

Rosé:

- Quartz Reef Rosé 2010: 100% Pinot Noir aged on lees for a minimum 24 months.
- Akarua Rosé Brut NV (Bannockburn): 74% Pinot Noir, 26% Chardonnay.
- Hans Herzog Cuvée Therese

TRANSFER METHOD: TRY THESE WINES

- Lindauer Special Reserve Blanc de Blanc: 100% Chardonnay
- Lindauer Classic Sec: normally 52% Chardonnay, 48% Pinot Noir with 24 g/L residual sugar.
- Lindauer Rosé

CHARMAT METHOD

- Millton Muskats @ Dawn
- Soljans Estate Fusion Sparkling Muscat

CARBONATION METHOD

- Yealands Sparkling Pinot Gris
- Villa Maria Private Bin Lightly Sparkling Sauvignon Blanc

Wine labels

Wine labels provide you with important information about the wine and its origins. Internationally, there has been a drive to standardise this information so that consumers are aware of what they are drinking irrespective of the country of origin. New Zealand has adopted the following 85% rule, so for instance when you see a wine labelled with a number of grape varieties, they will be appear in descending order of the proportion of the blend. (See p. 128).

MANDATORY REQUIREMENTS ON LABELS

1. Legibility; easy to read in English language.

2. Name and description of the wine: Greywacke Marlborough Pinot Noir 2012

3. Country of origin: 'product of New Zealand'.

4. Name and address of supplier/winery/producer.

5. Alcohol content declaration.

6. Net contents e.g. 750 millilitres (mL).

7. Standard drinks declaration: standard drinks (10 millilitres pure alcohol per litre of wine). *'contains approximately....'* The formula for calculating this is 0.789 x the actual content x the volume of the container in litres.

8. Warning and advisory statements: allergen declaration if wine contains more than 10 milligrams per kilogram of sulphur dioxide or contains preservative 220 (sulphites) or preservative 224 (potassium metabisulphite). Allergen declaration if milk, eggs, fish (except isinglass) are present.

9. Lot identification: all wines must bear a lot identification. If only one bottling run then this is not required.

Other requirements:

- Date of labelling is not required for bottled wine but may be required for wines of a shorter shelf life, e.g. bag in box wine.

- The 85% rule of vintage, variety, area included in the statement.

- Prohibited labelling on wine: cannot make health claims or false representations.

Words that may appear on labels:

- Appellation
- Vintage
- Estate bottled
- Superior
- Classic
- Old vines
- Reserve
- Winemaker's reserve
- Selection
- Bin

And some are just words from the marketing department...

THE 85% RULE

1. If the label states that the wine is from a single grape variety, vintage or area of origin, it must be at least 85% from that variety, vintage or area, e.g. a '2007' wine must contain at least 85% of vintage 2007 wine.

2. If the label states the wine is a blend of grape varieties, vintages or areas of origin, at least 85% of the blend must be from those varieties, vintages or areas, e.g. a 'Chardonnay Chenin Blanc' must contain at least 85% from Chardonnay and Chenin Blanc grapes.

3. If the label says the wine is a combination of grape variety, vintage, and area of origin, the combination referred to must be at least 85% of that wine, e.g. '2008 Marlborough Pinot Noir' must contain a minimum of 85% Pinot Noir from Marlborough that was harvested in 2008.

4. Blend statements must be presented in descending order of proportion in the blend, e.g. 'Chardonnay Chenin Blanc' must contain more Chardonnay than Chenin Blanc in the blend.

5. You cannot include a claim about grape variety, vintage or area of origin if the wine contains a greater percentage of wine from another grape variety, vintage or area not referred to on the label.

6. Export wine may make claims about grape variety, vintage or area of origin that differ from New Zealand's requirements when the overseas market access requirements (OMAR) have been notified under the Wine Act.

See http://www.foodsafety.govt.nz/elibrary/industry/meeting-grape-wine-labelling-requirements.pdf

Faults in wine

CORK TAINT

Cork remains a very popular closure, especially for the consumer, and recent innovations in cork manufacturing have seen the incidence of cork taint greatly reduce.

Smells: Dirty, dusty, mouldy, wet cardboard.

OXIDATION

Oxidation of wine occurs as soon as wine is exposed to air. Over time, the microbiological activity by bacteria and yeast will make the wine resemble vinegar (acetic acid) and become more brown in colour. Oxidised wine will start to lose its fresh, fruity aromas and develop toffee, caramel, Sherry notes that you would not expect to smell in that grape variety.

Smells: Loss of primary fruit, toffee, caramel, vinegar (acetic acid).

BRETTANOMYCES (BRETT)

Brett results from yeast in the winery and causes some feral smells. For many winemakers, it is regarded as a fault but for others it is considered to be a character of the wine. In New Zealand wines, we would not expect to smell these odours.

Smells: Barnyard, horse, even bandaid or vinyl.

OTHER SMELLABLE COMPOUNDS

These are generated by some bacteria and/or yeasts and can be a reason for judging a wine to be faulty, however, much comes down to individual taste. In many cases, the smells will just 'blow off' as the wine is aerated in decanter or glass.

- Mercaptans: smells of onion, garlic, boiled cabbage
- Volatile Acidity (VA) often found in sweet wines affected with botrytis: smells of nail polish remover, vinegar

Often these smells are described as 'reductive' or showing 'reduction'.

- Sulphur dioxide: smells of burnt match, fireworks
- Hydrogen sulphide: smell of rotten eggs

OTHER OBSERVATIONS

Tartrate crystals

Sometimes, in the bottom of a bottle of wine or on the end of a cork, you will find small crystals that may also look like tiny shards of glass. They are not a fault but are usually prevented by cold stabilisation before bottling.

Poor stabilisation

A wine that is hazy or cloudy may be the result of poor stabilisation that is caused by the growth of micro-organisms (yeast, bacteria) in the wine.

Bubbles

Bubbles in a still wine are the result of unplanned fermentation in the bottle. Sometimes, often after a still wine has just been bottled, you may feel a sensation of bubbles. This will disappear as the wine matures.

Ageing of wines

As wines age, they change in complexity but it is difficult to predict how the ageing process will evolve and how long it will take.

Wine must be intrinsically good to be capable of ageing. You cannot age a wine of poor quality and expect it to improve. Wines mature at different rates depending on the style and vintage characteristics. In red wines, the higher the levels of flavour compounds and phenolics (especially tannins), the longer it is capable of being aged.

When storing wines, the ambient temperature of the room will affects how a wine matures. The cooler the temperature, the slower the maturation; conversely, the warmer the temperature, the faster maturation proceeds. Wines do not like to be stored where they are subject to extremes of temperature or humidity.

FACTORS AFFECTING AGEING

- The quality of the wine: get advice from the winemaker, wine writer or specialist fine wine retailer about the ageing potential a wine.
- The temperature at which the wine is stored affects how the wine ages.
- Quality of the closure.
- Ullage refers to the air space between the wine and the closure. If the bottle is not filled to the correct level or wine has leaked out, it may be oxidised.
- pH level is important as in general the lower the wine's pH the longer the wine is capable of evolving.
- Sulphur dioxide concentration acts as a preservative in the wine.
- Residual sugar level.

AVERAGE AGEING PERIODS[3]

Whites

• Chardonnay	(1–6 years)
• Rieslings	(2–30 years)

Reds

• Pinot Noir	(2–8 years)
• Merlot	(2–12 years)
• Syrah	(4–16 years)
• Cabernet Sauvignon	(4–20 years)

Sweet

• Noble/Botrytised wines	(5-25 years)

NOTES:

1. This table is based on Robinson. J, (2006), p. 671.
2. Robinson, J. (2012), p. 1179.
3. This information is based on Robinson, J. (2006).

CHAPTER SIX:

Evaluating wine

For the new student of wine, it may seem that what is written by wine critics and taught by wine educators is both confusing and contradictory. There are so many new words, often with a range of meanings. However, with a little more knowledge, it is not difficult to grasp the fundamental concepts and build your confidence.

- Why do some combination of ingredients appeal and others not?
- Why can some tastes excite your palate and other repulse?

Sometimes it is hard to explain why you like to eat a certain dish or blend of ingredients but can't stand others.

Taste is a riddle

THE SERIOUS ASPECT

There is a whole science behind how we taste that underpins the food service industry and what we buy in the supermarkets – think of all those ready-made dips from pesto to hummus or wacky combinations of ice cream. Consumer focus groups evaluate and rate products to consider their appeal to future customers. Scientists also differentiate between taste as a chemical sensation received from our taste buds and flavour, which can be described as a 'fusion' of senses including:

Smell: identifying the aromas

Taste: What can you taste?

Tactile or Touch:

- **Chewy, crunchy:** You need to work on it, e.g. chewing meat, celery.
- **Soft, silky:** Think of the coating effect of chocolate, butter, cream (usually makes you happy).
- **Heat or burn:** Alcohol can burn and numb your mouth; drinking straight single malt whisky, Cognac.
- **Pain:** Eating hot chillies may cause pain for some people or excitement for others. Eating chillies or spicy food has the effect of leaving a numbing sensation in the mouth which can last for some time.

We learn to respond to taste. In most cases, we do not like food that is bitter; this is apparently a survival mechanism from our time as hunters and gatherers as bitter plants often contain poisons. Food that is too salty or acidic can also be unpalatable. Subconsciously, we are making decisions whether to swallow or spit!

On average, people have around 10,000 taste buds; some people may only have 1000 while others may have 13,000 and are highly sensitive to what they eat. The world of taste can look dramatically different depending on your physiology. When it comes to evaluating wine, having more taste buds does not make you better at tasting; it just means that you are more sensitive to taste.

There are also important psychological influences that are less predictable but equally powerful. We often associate a certain food or beverage with a happy memory - sitting on a beach or being on holiday in an exotic destination or just catching up with special friends; this too also influences our mood and how we perceive taste and flavour.

THE FUN ASPECT

Putting aside the science, it is fascinating to learn more about your own preferences and why you like certain things and dislike others.

It is also interesting to consider tastes or flavours that provoke behavioural responses, and this becomes a personal story of discovery. For me, I find that if I eat something sweet for breakfast – jam, chocolate spreads (not happening) or sweet pastries, this just sets me up to want to eat more sweet food throughout the day. If I eat a poached egg for breakfast, followed by espresso coffee, I often do not feel hungry until 3 pm. Yes, there are nutritional aspects here, but for me the coffee is bitter and also has a long savoury finish that I like. I do not want to alter the finish on my palate and so espresso is a 'full stop' and I don't feel that I want to eat for some time. What's more, if someone offered me a freshly baked lemon tart, just after I had finished the coffee, I would decline even if the tart looked amazing. The reason is that I do not want to introduce a citrus, sweet, slightly acidic flavour to my mellow coffee palate ... but this is me!

Building your palate preferences

DO YOU RECOGNISE THESE TASTES?

Learn to identify these tastes and rate the intensity for you on the palate. It is also interesting to consider whether you like a particular taste. I like to use the smiley face ☺ emoticon to remind me that this taste makes me happy! By the way, there is no right or wrong answer. It is about you and your preferences.

Sweetness

Eat white sugar

0 --- 1 --- 2 --- 3 --- 4 --- 5

low   high

Do you like to taste this flavour?

0 --- 1 --- 2 --- 3 --- 4 --- 5

hate   love

Sourness, acidity

Squeeze of lemon juice on your tongue

0 --- 1 --- 2 --- 3 --- 4 --- 5

low   high

Do you like to taste this flavour?

0 --- 1 --- 2 --- 3 --- 4 --- 5

hate   love

Umami, savouriness

Drink some miso soup, or taste Marmite

0 --- 1 --- 2 --- 3 --- 4 --- 5

low   high

Do you like to taste this flavour?

0 --- 1 --- 2 --- 3 --- 4 --- 5

hate   love

Bitterness

Eat grapefruit, lemon or orange pith or espresso coffee bean

0 --- 1 --- 2 --- 3 --- 4 --- 5

low   high

Do you like to taste this flavour?

0 --- 1 --- 2 --- 3 --- 4 --- 5

hate   love

Saltiness

Taste soy sauce or sea salt

0 --- 1 --- 2 --- 3 --- 4 --- 5

low   high

Do you like to taste this flavour?

0 --- 1 --- 2 --- 3 --- 4 --- 5

hate   love

Intensity of flavour

Life Saver Mint

0 --- 1 --- 2 --- 3 --- 4 --- 5

low   high

Do you like to taste this flavour?

0 --- 1 --- 2 --- 3 --- 4 --- 5

hate   love

The key is to learn to understand taste better and discover more about your palate and what combinations you like.

Vinotypes

Master of Wine and chef Tim Hanni has spent many years exploring taste sensitivity and has developed what he calls a person's 'vinotype'™. He describes a Vinotype as 'the combination of physiological and behavioural characteristics of a wine-drinking organism'. Hanni has devised four categories of tasters.

While some may find his approach curious, it is fresh and thought-provoking and provides an antidote to being told what are the 'correct' taste preferences that you should possess. It is presented here as a novel way of approaching the subject, which may help you discover your own taste sensitivity.

See more information at http://www.myVinotype.com

SWEET VINOTYPES

Are often acutely sensitive to light, sound, touch, smell and taste. They are often usually very artistic and tend towards the chaotic. You love fragrant, sweet wines that are typically low in alcohol and impeccably made. And they want sweet wines with the foods they love most, including steak or anything else!

HYPERSENSITIVE VINOTYPES

Are really sensitive to all sorts of things and, like the Sweet Vinotypes, they have many of the same pet peeves and pickiness. The big difference between Sweet and Hypersensitive Vinotypes is that Hypersensitive Vinotypes tend to prefer dry, or just off-dry wines on an everyday basis. Their favourite wines tend to be more delicate and very, very smooth while also being lower in alcohol. They may even like intense red wines, but not those with a lot of oak or heavy tannins; they prefer smooth and rich.

SENSITIVE VINOTYPES

Are more likely to go with the flow. They perhaps take coffee with milk or cream and/or a touch of sweetness at one point in the day, but black coffee if they feel like it. They are among the most adventurous wine lovers and open to all sorts of flavours and wine styles from delicate to robust. They do have limitations on bitterness and tannins; not looking for the oaky monsters but wines that are impeccably balanced, smooth and complex. Sensitive tasters are capable of registering relatively small differences or changes in sweetness, temperature or tactile sensations but are also a bit more flexible in adapting to the change.

TOLERANT VINOTYPES

Don't understand what all of the fuss is about with more sensitive Vinotypes. Intensity is the name of the game and the bigger, the better. As the antithesis of Sweet Vinotypes, they too want the wines they love, where and when they determine it is appropriate including with their food whether it is seafood, steak or salad. Just as long as it is red. Tolerant Vinotypes tend to be decisive and linear thinkers. More, bigger and faster usually equals better.

WHAT'S YOUR VINOTYPE?

18 to 25: Sweet

16 to 22: Hypersensitive

7 to 15: Sensitive

0 to 6: Tolerant

In the general population of wine drinkers, about 30% of people fall into the Sweet category, 25% into Hypersensitive, 25% Sensitive and 20% into the Tolerant group. Keep in mind that the more experienced and confident you are around wine, the more you will have already developed strong wine preferences through those personal experiences. You may look at your score and change your Vinotype to best fit the wines you love the most!

TIP! Just in case you run into a combination that makes your wine taste thin and bitter, try adding a tiny squeeze of fresh lemon juice and a pinch of salt to your food.

My Vinotype taste sensitivity Quiz™

This is a slightly amended version of Tim Hanni's 'My Vinotype' assessment quiz.

Circle the value that is CLOSEST to your preferences and then add up your score.

Gender
0   Male

3   Female

Salted snacks such as nuts and potato chips
0   I find most snacks too salty

1   Yay, I like salty snacks

3   Yum! I'm addicted to salty snacks

Salt preferences (answer from taste preference, not health standpoint)

0   I find many foods too salty

1   Food usually tastes fine as is, and/or I add a modest amount of salt when I cook OR I avoid salt for health reasons

2   I add a little extra salt to my food OR would like to but don't for health reasons

3   People give me a hard time for adding too much salt

Coffee or Tea
Describe the perfect cup of coffee or tea

0   I like it very strong (espresso or black tea, e.g. English Breakfast)

1   I like it strong (e.g. good espresso coffee or Earl Grey tea)

2   I like it medium (the weak coffee served at work, or green or herbal tea)

3   Coffee tastes so horrible I can't stand it

Sugar in your coffee
0   I drink coffee/tea with no sugar

1   A touch

2   One teaspoon or the equivalent

3   Two or more teaspoons

Cream or milk
0   I drink coffee black

1   Touch of cream or milk

2   Moderate cream or milk

3   Lots of cream or milk

Enjoy coffee with steamed milk or flavourings such as almond, vanilla, Irish Cream
0   No!

1   Cappuccino, latte, but NO flavourings

2   Sometimes

3   Yes

How do artificial sweeteners in diet fizzy drinks taste? (answer from taste preference, not health standpoint)

0   No taste problem (whether or not I choose to use them)

1   Don't know – never tried a diet fizzy drink in my life

1   Taste funny, but not too bad

2   I can tell a big difference but have adapted OR some are much better than others

3   Yuck! They taste horrible

Bonus question: Do you enjoy an occasional drink of straight Scotch, Cognac and/or Armagnac?
–3   Yes!

0   Sometimes

1   No way

Add up your points to get your score, which will determine into which of the four Vinotype groups you fall. Please refer to box on p. 134.

Formal wine tasting process

A wine differs from others in terms of its colour, texture, strength, structure, body, length and complex flavours.

'Assembling a vocabulary is a crucial element in becoming a wine connoisseur,' advises Jancis Robinson MW.[1] By using words to describe and identify the senses and flavours, it helps to clarify them and lets you evaluate a wine accurately.

AROMAS

1. Primary aromas: fruit and/or floral aromas of wine that originate from the grape.

2. Secondary aromas (bouquet): aromas of the wine that originate from the winemaking and fermentation processes, e.g. barrel fermentation (vanilla, spice); malolactic fermentation (nutty, buttery); and lees-stirring (yeasty, bready).

3. Tertiary aromas (bouquet): characteristics that develop as the wine ages, e.g. from extended ageing in oak barrels where the wine is exposed to oxygen (toasty, cedary); extended bottle age where the wine is not exposed to oxygen, (reductive notes: mushroom, cooked cabbage, tinned vegetables).

Olfactory sense: This refers to the smell and taste of the wine in the nostrils and behind the soft palate.

Flavours distinguished by the tongue: These include sweetness, saltiness, bitterness, acidity, and umami.

TASTING ENVIRONMENT

The tasting environment should ideally be as follows.

- Tasting room: odour free, with good natural light.
- Glasses: clean, odour-free glasses made of ISO-rated glass or similar; the glass should have a rounded bowl tapered upwards to capture the aromas as you swill.
- Spittoons: for spitting into after each tasting.
- White paper or tasting mat: to act as a background when reviewing the colour of the wine.

Prior to tasting, avoid eating strong foods or drinks (e.g. coffee, sports drinks) or brushing your teeth with toothpaste. Rinse your mouth with water to cleanse your palate; avoid being dehydrated.

SERVING TEMPERATURE OF WINE

Serve well-chilled 6–10°C
- Sparkling wines: Method traditional
- Sparkling Sauvignon Blanc, Muscat

Light to medium-bodied wines 7–10°C
- Light, fruity white wines
- Sauvignon Blanc, Pinot Gris, Riesling, Gewurztraminer
- Rosé

Medium to full-bodied white wines 10–13°C
- Chardonnay, Sémillon, Chenin Blanc,
- Barrel fermented Sauvignon Blanc

Light-bodied red wines 11–14°C
- Light-bodied Pinot Noir, Beaujolais

Medium to full-bodied wines 15–18°C
- Merlot/Cabernet Sauvignon blends
- Shiraz/Syrah
- Pinot Noir

Dessert wines Well chilled 6–8°C
- Late harvest, botrytised wines

TASTING PROCESS

1. Pour sufficient wine into the glass to be able to smell and swirl without spilling.

2. Hold the glass by the stem.

3. Smell once or twice before swirling to gain an initial impression. Concentrate. Your nose is good at sensing subtle flavours.

4. Tilt the glass to a 45-degree angle and look down at the clarity, colour and intensity of the wine.

5. Swirl the wine in the glass to aerate it.

6. Place your nose above the wine to smell the aromas. Sniff. What can you smell? Sniff again. Write down your impressions.

7. To taste, take a sip of the wine; swirl wine around mouth to cover the whole area of the tongue and expose all your taste buds and cheeks to the wine; hold; draw air over the top of your tongue; then spit out. Write down your impressions. A wine that has a persistent taste is often described as having a *long finish* or *good length*.

Where tastes are experienced

- **Sweetness:** Tip of the tongue
- **Acidity:** Edges, side of the tongue
- **Bitterness:** Back of tongue
- **Tannins (astringency):** Inside of the cheeks
- **Entrance to throat:** Burn of alcohol

Components of wine

Wine consists of a number of components that make an impression on your nose and palate but which also change and evolve during the tasting experience. This can prove frustrating, but with time you get used to the challenge!

APPEARANCE: HOW DOES THE WINE LOOK?

Action: Look at the glass, tip it slightly and look down on the wine

Clarity:
Clear – bright - dull - cloudy

Intensity/depth of colour from the core to the rim:
Pale – medium – deep

White wine:
Lemon green – lemon – gold – amber – brown

Red wine:
Purple (fuchsia) – ruby – garnet – tawny – brown

Rosé:
Pink – salmon – orange

Other observations:

- **Red wine:** In general, the deeper the colour, the younger the wine. All wines become brown with age.
- **Aged red wine:** Will lose colour at the rim, showing a 'brick' colour.
- **Viscosity:** Sugar or alcohol can make wine appear thicker and more viscous.
- **Tears/legs:** Streams of liquid that drip down the inside of glass after swirling.
- **Bubbles:** Fine or coarse.

NOSE: HOW DOES THE WINE SMELL?

Action: Smell the wine. First take a short sniff, then swirl the glass and sniff again.

Condition:
Smells fresh and clean; no obvious faults on nose

Intensity of smell:
Sensing the subtle aromas

Aroma:
Primary, secondary or tertiary

PALATE: HOW DOES THE WINE TASTE?

Action: Taste the wine, swirl it around your mouth.

Sweetness:
Dry, off-dry, medium, sweet

Acidity:
Low – medium – high

Body:
Light – medium – full-bodied

Tannin (reds, some whites):
Low – medium – high;
Ripe – rough – chewy

Alcohol:
Low – medium – high

Finish/Length:
How long does the flavour linger?
Short – medium – long

What flavours:
Fruits, oak, spice, earthy, herbaceous

How intense are these flavours?

Palate structure:

What sensations can you feel in your mouth?

- **Taste:** Sweetness, acidity, bitterness
- **Touch/Tactile:** Astringency, texture
- **Balance:** What is the relationship between sugar, fruit, acid, tannins and alcohol? Do they work well together?
- **Concentration:** Weak? Dilute? Intensely flavoured?
- **Complexity:** Lots of aroma, textures on the palate, and a long, lingering finish?
- **Identity:** Does the wine remind you of a place? Expressiveness?

Age of the Wine:

- **Too young:** Aggressive, harsh, not in balance.
- Drink now, not intended for ageing.
- Drink now but has ability to age for 2–3 years.
- Drink now but also has ability to age for many years.
- **Too old:** Amber, brick, oxidised.

'The glossier the colour and the more subtly shaded its different colour gradations, the better the wine,'
Jancis Robinson

NOTES:

1.   Robinson, J. (2013), p. 41.

CHAPTER SEVEN:

A short history of wine and hospitality

The wine industry today is very important to our economy as a producer, exporter and employer. It has become such an integral part of the New Zealand scene that it is easy to take it for granted, however, it is fascinating to look back and reflect on the curious twists and turns that have characterised its development.

New Zealand's temperate climate meant that fruits and vegetables could grow well, and a range of familiar varieties was introduced and grown. From the earliest European settlement, there was an element of the immigrant society that was familiar with French cuisine. Top-quality food had to be cooked according to French principles and would naturally need to be served with only the best French wines. Even in these early years there appeared a sensibility, especially among the small elite, that wine was fashionable, and while we tend to associate the nineteenth century as a period of high spirit and beer consumption, it is a revelation to learn of the extent of vineyard plantings in early New Zealand.

What is also interesting is that from our colonial beginnings, rules surrounding the supply and sale of alcohol were quickly introduced. Over the past 170 years, New Zealand's licensing laws have chopped and changed reflecting widespread concern about

The 1895 vintage under way at The Mission vineyard in Hawke's Bay.

public drunkenness, but we still do not appear to have solved this problem. It does make many people chuckle to learn that drinking wine with a meal in a restaurant, which didn't have a liquor licence, was still an illicit activity as late as the 1970s. Winemaker Danny Schuster tells the story of looking around Christchurch for somewhere to eat after his arrival in the country in 1971. He heard some 'clinking glasses' and managed to locate the source of the noise, found a small restaurant and promptly ordered dinner with a bottle of wine. *'Look sharp because I am hungry,'* he told the waiter. The wine arrived at the table, an Italian red, and with it, a candle. When he asked what the candle was for, he was instructed *'to put it in the bottle when cops come!'*[1]

Pioneers

From the time of the earliest European immigrants, grapes were planted in New Zealand. The first record of plantings is attributed to the Reverend Samuel Marsden, who had set up the Anglican missionary settlement in the Bay of Islands. During his seven visits from Sydney, encompassing the period from 1814 to 1837, Marsden was a great promoter of agriculture, and when the Kerikeri mission was established in 1819, he planted grapes.[2]

James Busby was appointed British Resident in the Bay of Islands for the period from 1833 to 1840. His role was essentially that of mediator of disputes between the local tribes and European settlers. In his spare time, Busby established a small farm with an extensive vegetable garden and vineyard adjacent to his home. We now know this building as the Treaty House at Waitangi.

Today, Busby is credited with establishing the Australian wine industry. By 1825, he had begun to experiment with grape growing on his 2000-acre Hunter Valley property.[3] Busby had studied viticulture and winemaking for a short time before immigrating to New South Wales with his parents. He wrote several pamphlets on winemaking, including: *A Treatise on the Culture of the Vine and the Art of Making Wine* (1825) and *The Manual of Plain Directions for Cultivating and Planting Vineyards*

and for Making Wine in New South Wales (1830); the latter was reprinted in New Zealand in 1862. Returning to Europe in 1831 to secure a colonial administrator role, Busby also made a four-month tour of vineyards in France and Spain and obtained cuttings from 570 varieties for both wine and fruit production along with another 74 cuttings that he had personally taken.[4] These cuttings included: *'Grenache – Black, good grape ...Hermitage – Fine black,'* which Busby called 'Schiraz'. He described the Muscat as *'white, good, oval, large; food flavour, two sorts.'*[5] The cuttings were eventually planted in Sydney's Botanic Gardens.

Over the next few years, Busby requested cuttings be sent to him in Waitangi; however, we do not know specifically what varieties he planted. Keith Stewart, in *Chancers and Visionaries,* argues that we should consider that Busby planted the vines that he was writing about – Syrah, from cuttings that he had taken from Hermitage; Marsanne, Roussanne, 'chaudeney, pineau noir' and grapes from Bordeaux. When Charles Darwin visited New Zealand in 1835, he noted that grapes were thriving in the Bay of Islands.[6]

The French explorer Dumont d'Urville visited Busby in 1840 and wrote in his log:

I was given a light white wine, very sparkling, and delicious to taste, which I enjoyed very much. Judging from this sample, I have no doubt that vines will be grown extensively all over...[7]

The foundations of a wine industry were therefore in place by the 1840s, although it took over 100 years for it to be truly established.

Henry Winkelmann (1912). Treaty House and Grounds.
Auckland War Memorial Museum – Tamaki Paenga Hira. PH-NEG-768

SETTLEMENT

Into Busby's isolated world, 1840 marked not only the negotiation and signing of the Treaty of Waitangi but also the arrival of the first British settlers in January at Port Nicholson (Wellington) and a rival French group who arrived in August at Akaroa and wished to claim the South Island for France. The French settlers also brought with them cuttings from grapevines.

The Roman Catholic Bishop, Frenchman Jean-Baptiste Pompallier, arrived in Hokianga in January 1838. By 1843, he had established a series of mission stations from Hokianga to Wellington in the North Island and at Akaroa and Otago in the South Island. Wherever the missions were established vines were planted, as it was common for the priests to drink wine with their meals. The most successful vineyards were those planted by the Marists in Hawke's Bay from 1851.

Edward Gibbon Wakefield, the British politician and director of the New Zealand Company, had been advocating a system of planned settlement since the 1820s. Wakefield believed that many of Britain's social problems were primarily caused by overpopulation and could be resolved by emigration to the colonies. His dream was to establish a microcosm of British society in New Zealand, attracting settlers from all classes – labourers, artisans and those with capital. The settlements were established in Wellington (January 1840), Wanganui (September 1840), New Plymouth (November 1841), Nelson (February 1842), the Church of Scotland settlement in Dunedin (1848) and the Church of England's Canterbury Association settlement (1850), regarded as the most successful. What welcomed these groups of colonists were rough and ready settlements, lacking any of the refinements of home, and usually at odds with what they had been led to believe existed. For many, this was deeply disturbing, and they felt defrauded by the advertisements and antics of the New Zealand Company.

Most of the new settlers eventually flourished in their new country and quickly sought to emulate the society of their prosperous British relatives. Faced with unsuitable public dining opportunities, the dinner party soon rose to be the ultimate dining experience for the well-off settlers. As in Britain, how they dined became an indicator of their financial and social success. Amy Trubek states that 'To dine well, the British had to dine French' and by the 1850s haute cuisine was *'de rigueur among the fashionable.'*[8] In New Zealand, French cuisine established a tentative hold among the elite, which was closely rivalled by Mrs Beeton, and her British guide to home management published in 1861.

Hotels and alcohol

New Zealand's first hospitality businesses were accommodation houses and hotels. In a colonial society that drew most of its members from Britain, it is understandable that the institutions of hospitality would reflect those of 'Home'. For an immigrant-based society it was essential to have some place to rent a bed upon arrival. The first hotels offered basic accommodation and simple food. Taverns historically provided lodging, meals and drinks; the 'grog' shop or pub, which was dominated by a counter or bar, primarily sold drinks.

As European settlement continued, hotels were required to be licensed to sell alcohol while accommodation houses could not sell liquor at all. Some of the first instances of lawlessness in New Zealand were as a consequence of drunkenness. In the 1830s, crews from whaling ships in the Bay of Islands were supplied with alcohol leading to much disorder in the absence of any police force. Alcohol use was prevalent among the settler communities. Historian Tony Simpson argues that the new immigrants were used to consuming high levels of spirits (whisky, gin and rum), beer and cider because the quality of water was unsafe in their homeland. Drunkenness was rife and any public event seemed inevitably to involve drinking to excess. In the period from 1845 to 1870, there were more than 15,000 cases of public drunkenness before the Auckland courts.[9]

DISTILLATION

The new immigrants had come to New Zealand from cultures where they were used to drinking spirits and the manufacture of spirits was often carried out at home. They brought this technology to New Zealand and promptly set up their own stills. Spirits were an inexpensive form of intoxication and played havoc with the local Maori communities not used to such 'fire water'. The first liquor law, the Distillation Prohibition Ordinance 1841, prohibited the distillation of spirits except by chemists for medicinal purposes. However, this was hard to police and private stills existed all over the country. The Licensing Ordinance of 1842 prevented the sale of liquor except by licensed operators. Both of these ordinances were issued in response to problems caused by intoxication.

The process of distillation goes back to Ancient Greece. Alchemists used the process to provide the base for many medicines, which were often referred to as aqua vitae – water of life. Arab scientists between AD 800 and 1000 had perfected the stills they used and refined the techniques.[10] These skills had returned to Europe over the subsequent centuries. Brandy first emerged as a medicine around 1300 and the first licence to manufacture was granted by Louis XII of France in 1514.[11]

For the English, gin was far more important than brandy. It had been introduced by William of Orange, a Dutch aristocrat who ascended the throne as joint sovereign with his British wife, Mary. Gin had its origins in Holland[12] and could be made from any grain. It took its distinctive flavour from the juniper berry and required no ageing. As early as 1725, there were 6187 gin shops in London.[13] Gin was consumed in epidemic proportions, resulting in wholesale drunkenness, and private stills were common. The Gin Act of 1736 was designed to curb this consumption; it raised taxes on retail sales of gin and introduced a prohibitive licence fee for sellers, effectively making the gin trade illegal.

In Scotland, whisky was the spirit of choice and was made from the distillation of what was essentially ale. By 1579, a series of Acts of the Scottish Parliament attempted to restrict production as distilling was widespread and a considerable domestic industry existed.[14] Illegal stills were the focus of subsequent governments in Scotland and Ireland. For example, in 1834 there were 692 seizures in Scotland and 8192 in

Ireland of illegal stills.[15] Rum also began to feature as an important spirit and was a daily ration for a naval man. At the establishment of a whaling station in Otago in 1831, the inventory included 'a pipe of gin and two puncheons of rum'.[16]

NEW ZEALAND WHISKY?

As wheat became more widely planted, it was argued that the manufacture of spirits was a 'simple and lucrative method of disposing of surplus grain'.[17] The Distillation Act was passed in 1868 and in a gesture designed to support the fledgling industry, the government applied only half the excise duty that was levied on imported spirits. With a distillery established in Dunedin and another in Auckland, a large quantity of 'home-grown' whisky poured on to the market at a much cheaper price than its imported competitors. This had an unanticipated effect, as the excise tax on liquor was an important source of revenue to all governments. The preferential rate given to the New Zealand product meant that the government was losing valuable income and by 1871 it warned that the preferential rate would need to be changed![18]

During this time, pressure to bring the sale of alcohol under popular control was mounting by the advocates of prohibition. In 1874, a parliamentary select committee considered the issues of distillation. The result was the Excise Duties Act 1874, which curiously adopted the policy of refusing distillers licences and forced the two existing distilleries to close. The government paid a total of £27,500 in compensation to these companies to close. Conrad Bollinger, in his book *Grog's Own Country*, argues that this turnaround may have had more to do with the Vogel Government's attempts to borrow money in London from the same financiers who backed the British spirit industry. At the time, Prime Minister Julius Vogel was raising loans in Britain to fund essential infrastructure such as road and railways. The preferential treatment given to the New Zealand spirit manufactures had damaged the trade of the powerful British whisky companies. It is difficult to avoid the conclusion that the enforced closing down of New Zealand distilleries was part of the price the

New Zealand Government had to pay for the millions it was borrowing.[19]

It was not until 1958 that distillation again commenced in New Zealand and, even then, this was only with experimental licences.

Fortified wines such as Port and Sherry were popular, but they required the addition of a spirit to fortify them to the appropriate level. With distillation forbidden, the wineries had to use imported grape spirits, which was costly, and this policy caused much frustration among winemakers. Around 1890, after much lobbying, legislation was passed allowing vineyards of two acres or more to install a still in order to make the spirits for fortification.[20] In 1908 this was amended to allow vineyards of five acres to distil their own spirit for fortifying wines produced on their vineyard.[21]

SUCCESSFUL GROWERS

By the 1890s, grapes were being grown all over the country. Charles Levet and his son William established New Zealand's first commercial vineyard in Kaipara, near Auckland, in 1863. They made fortified wines as well as fruit wines. By 1895 the estate was known as Lord Glasgow's vineyard, reflecting the popularity of the wine with the current governor, who let Levet use his name. Spanish winemaker José Solé, who changed his name to Joseph Soler, planted vines in Wanganui in 1866 and by 1880 had won six awards at the Melbourne International Exhibition. Soler went on to make sparkling wine and planted grapes such as Gewurztraminer and Pinot Noir. The Mission, now located at Meeanee, benefited from the arrival in 1871 of Brother Cyprian Huchet who was from a winemaking family in the Loire Valley. He went on to reform the viticulture, adding Pinot Noir and Pinot Gris, extending the vineyards and improving wine quality.[22]

In Marlborough, David Herd planted vines at Auntsfield in 1873. He was managing Meadowbank Station and planted a one-acre plot of Muscat. Old photos of the early vineyard show the vines were trailed over loose wires, held up by tall manuka poles.[23] Herd bought part of the station in 1879 and

continued to make wine until his death in 1905. Bill Paynter, Herd's son-in-law, then took over and continued making wine until 1931.

The restored cellar at Auntsfield Estate, Marlborough, built in 1873, features a rammed earth floor and manuka pole roof.

Station owner William Beetham, influenced by his French wife, established a small vineyard at his home in Masterton in 1883. He was so impressed that he planted more vines at Lansdowne Station. This became a commercial winery producing 8500 litres of wine annually.[24] In turn, Beetham encouraged other wealthy landowners to consider vines. Inspired by Beetham, Henry Tiffen from Greenmeadows Station, Taradale had 12 hectares of vineyard, primarily of Pinot Noir and Pinot Meunier, planted by 1896. With an annual production of 10,000 litres, fermented in 1000-litre totara vessels, the wine was pressed and aged for three years before release.

In 1892, Bernard Chambers from Te Mata Station planted Pinot Noir and Black Hamburg with cuttings obtained from the Marist Brothers. By 1911, there were 14 hectares planted on the slopes of Te Mata Peak with grapes including Pinot Meunier, Syrah, Cabernet Sauvignon, Riesling and Verdelho producing 54,000 litres of wine.[25]

Israel Wendel and his daughter Brunetta had established a vineyard in central Auckland by 1893. They also owned a wine bar on Karangahape Road and wine cellars in Symonds Street. Brunetta is considered New Zealand's first female winemaker.[26]

In 1902 Assid Abraham (A.A.) Corban purchased ten acres in Henderson and started to plant grapes.

He was soon followed by a long line of Dalmatian families. By 1960, Corbans was New Zealand's largest winery.

Khaleel, Zarefy and Annisie carry the message to Taranaki.
Auckland War Memorial Museum – Tamaki Paenga Hira.
PH-CNEG-c26512

Romeo Bragato

At the invitation of the New Zealand Government, Australian-based viticulturist Romeo Bragato visited New Zealand in 1895 and was able to see grapes and taste wine from Arrowtown, Cromwell, Akaroa, Moutere, Wairarapa, Hawke's Bay, Bay of Plenty, Auckland and Northland. In Akaroa he tasted ripe grapes of the Chasselas, La Folle and Muscat varieties.[27] In the Wairarapa, he visited Beetham. In Hawke's Bay he visited the Marist brothers at The Mission, Tiffen and Chambers at Te Mata and also visited the vineyards in Auckland.

Bragato recognised that viticulture was viable in New Zealand; land was cheap and only small plots were needed in order to produce a reasonable quantity of wine. He even recommended that the government plant oak forests in preparation for 'a large wine-producing colony'.[28] From these records, there is a sense of a fledgling but vibrant wine industry. Throughout the country he encouraged the establishment of local vine growers' associations. For its part, the government, through the Department of Agriculture, was active in promoting viticulture.

For the New Zealand conditions, Bragato recommended the following *Vitis vinifera* grape varieties: Black Hermitage (Shiraz), Cabernet Sauvignon, Cabernet Franc, Dolcetto, Pinot Noir, Pinot Meunier, Riesling, Pinot Blanc, Tokay, and White Hermitage (La Folle, Marsanne, Clairette Blanc).[29]

On Mt Eden Road, Bragato found two vineyards with over 100 vines infected with the sap-sucking insect phylloxera. He had seen the damage that the aphid-like insect had inflicted on European vines and so recommended instant destruction, but the growers were reluctant to destroy their investment. It was not until the government intervened with legislation that inspectors began enforcing destruction.[30] By 1899, over 100 infested properties had been identified, mostly in Auckland, but also in Wellington and Hawke's Bay.[31] Further outbreaks in 1901 resulted in Bragato's return to New Zealand and subsequent appointment as the Government Viticulturist.

Based at Te Kauwhata, which had been established as a viticultural station in 1897, Bragato supplied growers with phylloxera resistant vines that had been grafted on to American rootstock. He taught the growers advanced viticultural techniques and also experimented with vines from all over Europe to evaluate their performance in New Zealand conditions.

At the time there appeared to be widespread adulteration of wine with chalk and other products. Bragato therefore recommended to the government that winemakers be licensed and their products tested. In 1906, Bragato published *Viticulture in New Zealand,* which sold 5000 copies and explained techniques such as grafting onto phylloxera-resistant American vines. In 1908, five wines from the Te Kauwhata Experimental Station won gold medals at the Franco-British wine exhibition held in London.

The changing political environment and the influence of the prohibition movement in New Zealand, however, troubled Bragato. Increasingly, his requests for resources fell on deaf ears. When his position was restructured, he resigned and left New Zealand in 1909. By this time there were over 269 hectares of vines planted in New Zealand.[32]

Prohibition and hospitality

From our perspective in the twenty-first century, the notion of prohibiting the sale and consumption of alcohol seems a curious phenomenon. From the early nineteenth century, concern about widespread intoxication and the social problems arising from grog shops and beer shops had been increasing in the United States, Britain and its colonies. Those supporting prohibition came from a variety of backgrounds and classes.

Prohibition or what became known as the Temperance Movement also embraced progressive politics such as the call for women's franchise and property rights that would enable women to vote and own property equally with men. The Methodists and Presbyterians were prominent in the Temperance Movement but so too were the Salvation Army, Anglicans and Roman Catholics. Clergy often took leadership roles, as did many women including Kate Sheppard, who today remains the acknowledged leader of this time. Central to their principles was the belief that society would be better served by giving women the vote, because women were viewed as the key to overcoming the abuse of alcohol in the home and in public.

The Women's Christian Temperance Union was established in 1885 and the New Zealand Alliance for the Suppression and Prohibition of the Liquor Traffic was formed in 1886 to provide a united front for the temperance movement. The president was former Prime Minister Sir William Fox and the 16 vice-presidents included the Chief Justice, Sir Robert Stout. Prohibition was also strongly linked to the labour movement that emerged in the 1880s and 1890s in New Zealand. In 1893, New Zealand women were the first women in the world given the right to vote.

Kate Sheppard, leading women's suffrage advocate.
Alexander Turnbull Library, Wellington, New Zealand.
Ref: 1/2-C-09028-F.

LICENSING

The Licensing Act of 1881 established a raft of regulation that essentially remained unchanged until the 1960s. It established licensing districts and licensing committees composed of a resident magistrate and five elected members who were ratepayers from that area. These committees remained in place until 1961. Initially, elections were held annually but, in 1889, the Triennial Licensing Committees Act allowed for three-yearly elections.[33] This Act required pubs to close each night at ten o'clock and they could not open again until six o'clock the next morning. Hotels were also prevented from opening on Sundays and this would not be changed until 1989. In an interesting twist, the Sale and Supply of Alcohol Act 2012 has reintroduced the district licensing committees with the goal of communities becoming more active in management of the sale of alcohol.

LOCAL OPTION

As the campaign for prohibition gained momentum, new legislation was passed in 1893 that transferred licensing policy in a region to the ballot box. People were now able to vote directly on the issue of the sale of liquor in their area. This local-option poll was to be held at the same time as the parliamentary elections. Licensing districts now corresponded with the electorate seats. With regard to licensed premises, residents could vote for:

- Whether the present number could continue.

- Whether the number should be reduced.

- Whether any licences should be granted at all; that is, no licence or 'dry'.

GOING DRY

The Clutha electorate in Central Otago became the first electorate to go dry in 1894, after the general election of 1893, the first in which women could vote. Over the next decade, there were amendments to refine the detail of the legislation. A new Licensing Act in 1908 and a further Licensing Amendment Act in 1910 introduced a national referendum on prohibition to be held every four years. Prohibition would be introduced to New Zealand if the referendum obtained a three-fifths majority of the voting population. The first referendum was held in 1911. The 1910 Act also prevented the employment of women in bars, except for the family members of the licensee and those barmaids employed prior to the change in legislation. This was not repealed until 1961. At the same time the minimum drinking age was raised to 21 years.

The prohibition movement was achieving success in both legislation and on the ground, in the electorate. Mataura (Southland) and Ashburton had gone dry in 1902, Grey Lynn, Invercargill and Bruce (Otago) in 1905. Masterton, two Wellington electorates and Eden had gone dry in 1908. In Invercargill, the pubs closed in 1906 for 40 years! In 1908, the Eden electorate, which included parts of the winemaking region of Henderson, had voted for no-licence.[34] By 1910, 12 of the 76 general electorates were dry and licences had dropped from 1719 in 1894 to 1257.[35] By 1967, there were only 1200 licences for a population of around 2.75 million.[36]

Opposition to the changes in licensing resulted in the formation of the Wellington Licensed Victuallers Association in 1902. This organisation would be known as the Hotel Association of New Zealand (HANZ) from 1958 and more recently (1995) the Hospitality Association of New Zealand and Hospitality New Zealand (2011). The hotel owners, like the clubs, the brewers and the winegrowers, looked with horror at the undermining of their industry. It had become a political lottery where every three years your electorate could vote for prohibition and therefore force all licensed premises to close.

Corbans Original Wine Shop. Corban's winery was located just inside the boundary of a no-licence area. After 1908, they built a depot in the neighbouring electorate that was 'wet'.
Auckland War Memorial Museum – Tamaki Paenga Hira.
PH-CNEG-c21523

Under the 1881 legislation, social clubs selling liquor were required to obtain a 'charter' from the Colonial Secretary and submit details of their organisation on an annual basis. Prior to this, clubs had not been affected by any legislation and members took offence at the prospect of police supervision. Clubs could only sell liquor to members or their guests. Under the Licensing Amendment Act 1904 the position of clubs was clarified and essentially put on the same footing as hotels. They too were prevented from selling liquor after ten o'clock at night and on Sundays, although liquor could still be sold to members staying in club accommodation. The Act also banned the playing of cards or billiards for money. If the electorate voted for no licence the club was forced to close or continue but without selling alcohol. On legal advice, some clubs evolved a 'locker' system where members kept their own alcohol on site for their own use.[37]

Sparrow Industrial Pictures Ltd. Glass of Ale.
Auckland War Memorial Museum – Tamaki Paenga Hira.
PH-NEG-SP-1698dm

SIX O'CLOCK CLOSING

With the onset of the First World War, the prohibition movement continued to lobby the government. The Sale of Liquor Restriction Act was passed in 1917 and forced hotels to be closed at night from six o'clock until nine o'clock the next morning. Six o'clock closing, with its 'six o'clock swill' or binge drinking before closing time, was introduced as a temporary wartime efficiency measure, but it remained in force until 1967. A loophole emerged where hotels or chartered clubs that served substantial meals could still sell alcohol from six o'clock until eight o'clock. Resident guests were able to ignore the closing time.

REFERENDA

In 1918 legislation was introduced to hold a special poll on national prohibition before April 1919. In this poll there were only two options: continuance or prohibition with compensation for the businesses affected. The prohibitionists saw this referendum as the culmination of decades of work. Bill Brien, in his history of the Hospitality Association, writes:

> The prohibitionists campaigned fervently, but a leading Auckland brewer, Moss Davis, at his own expense went to London to win the support of the soldiers to vote for continuance. The trip paid off handsomely. The soldier's vote kept the country wet and while there was a majority for national prohibition in New Zealand, the votes of the soldiers in the Expeditionary Force resulted in a 10,362 majority for national continuance.[38]

The result was that 51 per cent voted for continuance in the poll. This was a very narrow margin as only a vote of 50 per cent was required for New Zealand to adopt prohibition. This outcome was very disappointing for the prohibition organisers. Some felt that the payment of compensation to those businesses that would have to close affected how people voted:

> ... there is no doubt that a great number of individuals voted solely against a payment of four and a half million pounds of public funds. Many confirmed prohibitionists felt so strongly that such a payment was wrong that they declared they would never vote in favour of the proposal.[39]

In 1919, the United States introduced prohibition that was to last until 1933. In New Zealand, a second poll was held in December 1919, to consider three issues: continuance; state purchase and control rather than the controversial 'compensation'; and prohibition. This poll required that the prohibition vote needed to be greater than the combined totals of continuance and state purchase. In this ballot, the result was even closer with a small margin of just over 2000 votes against national prohibition. However, this was the turning point, as through the 1920s, the drive towards prohibition lost ground. In the 1925 referendum over 36,177 people voted against prohibition. At this time the Ohinemuri electorate in the Coromandel area was the first electorate to vote for 'restoration' of its liquor licensing after 17 years of no licence.

Robin Morrison (1979). Fish and chip shop, Kaitangata.
Auckland War Memorial Museum – Tamaki Paenga Hira.
PH-NEG-RMS-FTR103-180.

RECUPERATION

In the years that followed, most electorates abandoned prohibition, however, there remained a complex web of regulation. Many hoteliers, who had removed bar areas when their electorate went dry, replacing them with accommodation, required new capital investment to convert their hotels back when the electorate became wet again. Into the breach stepped the breweries and there was significant rivalry among them to contract hotels. In 1925, New Zealand Breweries organised a scheme where it would purchase a hotel outright and then employ managers or lease back the property to the former owner. It would then impose a strict 'trade tie and a rental' agreement that forbade the sale of rival brands of beer.[40] Dominion Breweries was established in 1930 and challenged the 'tied' house system of New Zealand Breweries with its own version. Conrad

Bollinger describes the pub of the day as 'a long bare room with cold hard walls and floor, rather like a public lavatory, to enable it to be easily cleaned out with a hose'. Regulations required that passers-by could not look in to see people drinking so windows were often installed to prevent customers seeing out and others looking in. Later on carpet was introduced and dim lighting helped to conceal the dirt. The room was 'devoid of furniture except for the long elliptical bar in the centre'.[41]

MAKING YOUR OWN FUN

Until the 1970s, middle-class men would not go to the pub and it was certainly off limits for women of reputation. The private dinner party remained a popular social activity, especially in the period before restaurants were common. For many, 'the club' offered an important alternative for socialising.

There were clubs for everything and by this time women were well represented in the memberships: mah-jong, bridge, golf, cricket, tennis, bowls, rugby, sailing, Rotary and politics, to name but a few. My grandparents belonged to the English Speaking Union and I remember quizzing them about what they actually did at this club. As with so many clubs, the purpose was obscured by the value of the social contact.

The Christchurch Club was established in 1856, only six years after the city had been founded. It was exclusively for men until 1969 when women were introduced as associate members. My family's Uncle Jim[42] told me about his introductory dinner at the Christchurch Club in 1938. He was 22 years of age. It was a black-tie affair with a special dinner selected for the occasion by a senior Hay relation. They were a party of four and dinner started with martinis in the smoking room. The butler announced dinner and they went into the dining room. The first course was soup served with sherry; the fish followed with a bottle of Pouilly Fuissé (Chardonnay from Burgundy); roast beef with Yorkshire pudding was served with Château Mouton Rothschild. For pudding, crème brûlée was served with a sweet wine such as a Barsac or Sauternes. The savoury course was marrow served on silver spoons, which Jim described as a *'quivering mass of fat and veins'* and accompanied with a spirit that took away the fatty taste. The fruit course was next and '... *you watched to see how they worked it. I chose a banana because it was easy to peel with a knife and fork.'* Port followed and at this point the loyal toast was given: a toast to the King. Cognac and cigars were taken in the smoking room. This process took around one and a half hours; *'by then we had steam coming out of the top of our heads!'*

DALMATIAN INFLUENCE

From the 1860s onwards, New Zealand received numerous immigrants from Dalmatia, a region of present-day Croatia. Many found work on the goldfields or the gumfields. Josip Babich made his first wine, as a 16 year old, in the gumfields of Northland. In 1919, he purchased 24 hectares near Henderson where he raised cattle and grew vegetables, fruit

trees and grapevines. An influx of other Dalmatian families followed over the next 30 years – Nobilo, Brajkovich (Kumeu River), Soljan, Erceg (Pacific Vineyards), Delegat, Yukich (Montana), and Fistonich (Villa Maria) to name but a few. They grew fruit and vegetables to support and sustain their families, and it was the second generation who went on to be commercially focused.

The first Labour Government introduced import licences in 1935 and increased levies on imported wines, which made local wine more affordable and attractive. There remained however, some fundamental issues of quality that needed to be addressed and which were in part a consequence of New Zealand's curious liquor laws. The quality of grape varieties that were grown and their ripeness at harvest remained a problem. Sugar was commonly added as well as water to dilute the high acid from unripe grapes. Peter Babich commented in 2012:

> When I began working with New Zealand wine in 1948, it was all wrong. ...The law around selling wine in a minimum of 2 gallons was wrong, the food and drug regulations were wrong. How on earth could you have an area where grapes ...were allowed to be grown, but the wines weren't allowed to be consumed? ...The grape varieties themselves were wrong, the style of wine we were making was all wrong. [43]

In 1949, Frank Berrysmith was appointed government viticulturist and shortly after, Denis Kasza, a graduate in plant physiology from Budapest with a postgraduate degree from Montpellier University, also started working at Te Kauwhata. Kasza quickly identified that most wineries were harvesting grapes too early, resulting in highly acidic wines that required the addition of sugar to balance the acidity and achieve a certain alcohol level. Kasza went on to analyse different grape varieties and examine how to improve them.

Towards a
modern industry

MURMURS OF CHANGE

In 1945 a Royal Commission on liquor licensing was set up to review the industry. The Commission sat for 19 months and travelled throughout New Zealand receiving submissions. Opinions remained wildly divided on this issue and the proceedings were recorded in 52 volumes with a total of 7824 pages.[44] Even the breweries were not necessarily advocates of reform. With the legislative restriction on new liquor licences being issued, existing licensees had a monopoly to sell alcohol in their area. Interestingly, the breweries and licensees did not support extra trading hours because evening trade was subject to penal (overtime) rates for wages. The advantage of six o'clock closing if you were a publican was that you got the night off. The same amount of beer was drunk in the rush before closing, but the business overheads were far less. Bollinger, comments: 'All these laws were regarded, when they made the statute books, as mighty victories for the prohibition movement. But every one of them has turned out to the enormous advantage of the publican – or brewer who is so often his actual or virtual employer.'[45]

In its recommendations to Parliament, the Commission advocated that the government purchase all the breweries and place them in a public corporation. Not surprisingly, this did not happen. The Licensing Amendment Act 1948 was the result of the Commission's work. It established the Licensing Control Commission to review all existing licences and introduced a new form of licence known as the licensing trust. This format was often used when areas were 'restored' and was a type of public ownership where the trust would own and control all the licensed premises within an electorate. The other intriguing recommendation from the Commission was that public bars should be able to have seats!

In 1957, a parliamentary select committee recommended that single-bottle sales could be made from wineries and that wine could be sold in restaurants. Prior to this, winemakers had been prevented from selling single bottles of wine. By 1963, New Zealand had started to export wine.

Robin Morrison (1979). Royal Hotel, Naseby.
Auckland War Memorial Museum – Tamaki Paenga Hira.
PH-NEG-RMS-rFTR134-239

GROWTH

The 1960s and 1970s were a period of enormous growth for the wine and hospitality industries. In 1960, just 10 restaurants were licensed throughout the country.[46] The number increased slowly over the years. The repeal of six o'clock closing in 1967 was the watershed and marked the beginning of a new era for hospitality businesses. Instead of drinking at home or illegally behind blackened curtains in a pub, people were able to go out at night. In retrospect, it is hard to comprehend the enormous social change that has occurred in response to something as mundane as an alteration to the sale of liquor laws.

Alex Corban became New Zealand's first qualified winemaker, having trained at Roseworthy College in Australia. He engaged in a great deal of experimentation. He pioneered the German technique of *susse reserve*, known as back-blending in New Zealand, where sterile grape juice was added to a finished wine, which had the effect of making the wine sweet, fruitier and lowering the alcohol. In 1961 Montana released Pearl, a neutral wine served in a bulb-shaped bottle with screw closure, followed by Cold Duck, a red wine flavoured with fruit essence that was fizzy and sweet.[47] Corbans released Premier Cuvée, a tank-fermented sparkling wine with some Muller-Thurgau.

Australian vintner McWilliams bought the McDonald winery in 1962 with Penfolds bought the Averill Brothers Lincoln Road winery in 1963. Both companies were closely associated with breweries; McWilliams was a minority shareholder in a company that included Dominion Breweries and New Zealand Breweries. Dominion Breweries owned 40 per cent of Penfolds New Zealand. Keith Stewart argues that both companies helped improve the quality of winemaking because of their commercial experience in Australia.[48]

McWilliams went on to enjoy great success with Cresta Doré (white), Bakano (red) and Marque Vue (sparkling). In 1961, George Fistonich established Villa Maria in Mangere. In 1963, Corbans restructured to raise capital by selling down the family shareholding to a group of liquor merchants who held 19 per cent of the company. This provided capital for expansion into Gisborne and to upgrade their winery in Henderson.

Cooks New Zealand Wine Company was set up in 1969 by a group of Auckland businessmen and eventually floated on the New Zealand share market. By 1977 sales of Cook's Chasseur provided 80 per cent of their turnover and there were not enough grapes to meet demand.

MULLER-THURGAU

During the 1960s many hybrid vines were replaced with *Vitis vinifera* vines of which the German crossing, Muller-Thurgau, often called Riesling Sylvaner, was the grape of choice. It offered delicate fruit and floral aromas, and was easy to grow, with large yields. In the Gisborne region an increasing numbers of farmers converted parts of their farms to vines and became contract growers to the wineries. This model of contract growing was an important method of funding expansion as the wineries did not have to pay for the land and the sheep and cattle farmers welcomed the diversification.

Nobilo's Muller-Thurgau was first released in 1975 and was essentially a copy of the German Liebfraumilch style that was very successful internationally. Nick Nobilo comments:

I analysed the wine, saw what the alcohol, pH, sweetness and acidity were and discovered why it tasted that way. It was only 9.5 per cent alcohol, which meant it had been back-blended. This meant it had unfermented grape juice put into the finished wine to add weight, flavour and to reduce the alcohol content.[49]

This wine was an overnight success, attracting sweet Sherry and Port drinkers. Nobilo's was also instrumental in establishing Pinotage as an easy-drinking red wine.

In 1988 Nobilo's launched White Cloud, which consisted chiefly of Muller-Thurgau, 25 per cent Sauvignon Blanc with 15 per cent back-blend of unfermented Muller and some Muscat. White Cloud was originally designed as an export wine to develop new opportunities offshore and featured New Zealand imagery to evoke a sense of purity: 'land of the long white cloud'. It was initially sold to the state liquor monopoly in Sweden, shipped in tanks and bottled in Sweden. At one point, White Cloud made up 10 per cent of New Zealand white-wine production.[50]

ENTER SAUVIGNON BLANC

Sauvignon Blanc had intrigued Ross Spence, from Matua Vineyards, while he was studying at the University of California (Fresno) in the mid-1960s.

This wildly aromatic variety which ripened mid-season caught my attention. I remember thinking, an early to mid-ripening grape like this one was just the thing for a cool climate like New Zealand's and, once I was home in the late 1960s, I came across a Sauvignon Blanc clone at the Te Kauwhata Viticultural Research Centre. There wasn't enough to plant an entire vineyard so I grafted the vines I could get onto 1202 rootstock until I had sufficient to plant 250 vines.

While these did not flourish, in 1970 he located a new clone in an experimental block that had been planted by Corbans in Kumeu.

After much searching with an ampelography [grapevine identification] book in one hand and notebook in the other, I found one Sauvignon Blanc vine and a Sémillon vine alongside it. ...To the best of my knowledge, nobody knew about the new import prior to me finding it.[51]

These vines would provide the cuttings that were used by Montana and others to establish the Marlborough vineyards on the Wairau River plain in 1973.

GROWTH OF RESTAURANTS AND CAFÉS

Yet another Royal Commission into the liquor industry was held in 1973 and again looked at hours of trade and licensing. One issue canvassed in the submission from the Auckland Leagues Club was that *'sociologists were warning that many people would have difficulty in coping with large amounts of leisure time...'* and chartered clubs should look to incorporate poker machines as *'an amenity'*.[52] This was a portent of things to come.

Philip Temple, Orsini's proprietor, presenting a bottle of wine to a customer - Photograph taken by Ray Pigney.
Alexander Turnbull Library, Wellington, New Zealand.
Ref: EP/1988/4458/2.

During this time more restaurants and cafés were established. They were often run by European immigrants who may not have had a hospitality background, but their culture had instilled in them a deep appreciation of hospitality. The introduction of bring your own (BYO) wine introduced many more people to the pleasures of wine when dining out, and people were drinking New Zealand wine: Cold Duck, Blenheimer and Cook's Chasseur had become household names. In 1979, George Fistonich opened the first winery restaurant at Vidal's in Hawke's Bay.

OVERCOMING QUALITY PROBLEMS

The wine industry had become hugely competitive but still quality was undermined by the use of excessive amounts of water and sugar during winemaking. The Wine Institute of New Zealand was established in 1975, funded by an annual levy based on production. It was the first time that all the wineries had combined to work together to improve and move the industry forward. The water-in-wine practice undermined the creditability of the industry. In 1980 the government moved to make it illegal to add water during winemaking and in 1981 changed the sales tax and freed up the importation of wine.

By 1981, the Winemakers Act was passed and new regulations were introduced for the export of wine. In 1982, the first tasting of New Zealand wines was conducted in London, in what was to become an annual event, and increased the profile of New Zealand wines in that market.

Further tax increases in 1984 and the lifting of import duties on wine under the fourth Labour Government were the beginning of a new era of change. By 1984 Cooks had merged with McWilliams and was controlled by Brierley Investments. A large vintage in 1985, where 78,000 tonnes were crushed while the demand was only 55,000 tonnes, resulted in a price war. Price cutting followed with Villa Maria, Glenvale, Nobilo and Delegat's all finding themselves on the cusp of financial collapse.

In 1985 the major companies, Brierley (Cooks-McWilliams), Corbans (with Rothmans a major stakeholder), Lion Breweries (Penfolds), Montana, and Villa Maria had managed to lobby the new Labour Government to fund the vine-pull scheme. In 1986, at a time when the same government was cutting agricultural subsidies to sheep farmers, grape-growers were offered $6175 per hectare to pull 25% of the national vineyards. As a result 1515 hectares of vines were pulled out, primarily Muller-Thurgau. While the intention was not to replant grapes, many growers did use the payout to replace their vines with the more fashionable varieties of Chardonnay and Sauvignon Blanc.[53]

Further restructuring of the industry followed with Montana acquiring Penfolds New Zealand in 1986 to control 40% of the domestic wine market. Corbans bought the winery and assets of Cooks-McWilliams in 1987; this equated to around 30% of the market. Villa Maria went into receivership for a short period but was able to trade its way out and in 1987 bought the Glenvale winery in Hawke's Bay, which included Esk Valley.

Wineries that were focused on top-quality winemaking had started to emerge during this same period. Danny Schuster's 1982 St Helena Pinot Noir created waves as people could see the possibilities of this grape. Te Mata released its first Coleraine and Awatea in 1984 using classic Bordeaux grape varieties and winemaking techniques. David Hohnen established Cloudy Bay. Daniel Le Brun, originally from Champagne, also settled in Marlborough. In 1982 at Rippon Estate in Wanaka, Rolfe and Lois Mills planted the first block of vines with the aim of growing grapes to make high-quality wine.

Large vintages from 1989 encouraged the winemakers to export their surplus. Sauvignon Blanc was the star. With international recognition growing for the wines, more emphasis started to be placed on ensuring a coordinated marketing strategy.

By 1993, the industry leaders were talking about the necessity of planting more vines in order to reach sufficient annual stock levels to ensure a viable export industry. This goal set in motion a period of enormous investment in the wine industry. International as well as local investors poured in millions of dollars. Demand for local wine was also fuelled by publicity and international recognition of the growing sector.

LIBERATION

The Sale of Liquor Act 1989 was a radical departure from the past. Finally, the old regime was changed and if you were of good character, complied with the local government zoning policies and fulfilled the regulations governing environmental health, you could apply for a liquor licence. Overnight, numerous restaurants and cafés applied to sell alcohol. In turn they needed to put together a wine list. Most operators had little knowledge of wine and felt particularly nervous when faced with all those unpronounceable words on the labels of French wine. The fledgling domestic wine industry was a far more comfortable choice; after all, an element of parochialism can be attractive to the consumer.

The wineries as well wanted to show off their wines by providing on-site tasting rooms. Under the new legislation, for wine to be consumed on the premises, food needed to be available. This requirement led to the establishment of winery cafés and restaurants throughout New Zealand's wine country. The investment of substantial amounts of money to build magnificent winery restaurants, run by professional staff, has introduced a completely new experience to the hospitality industry.

To this happy blend of new bars, cafés, restaurants and wineries comes another important influence: the establishment of New Zealand as a first-class tourist destination. The surge in international visitors during the 1990s and 2000s has driven significant investment in the hospitality industry. The development of high-quality boutique lodges has created another specialised market that did not exist before. Better hotels and improved conference facilities all help make New Zealand attractive for international tourists. And tourists must be entertained with compelling and appealing activities, like wine tasting and visiting picturesque vineyards with gourmet restaurants and tasting delicious local cuisine.

NOTES:

1. Personal communications with Danny Schuster, 2006.
2. Dictionary of New Zealand Biography, Vol. 1 (1990), p. 272.
3. Ibid., p. 61.
4. Stewart (2010), p. 23.
5. Busby, James (1832), *Vines Introduced into the Colony of New South Wales.*
6. *Dictionary of New Zealand Biography,* Vol.1, p. 61.
7. Quoted in Scott, p. 16.
8. Quoted in Trubek (2000), p. 42.
9. Quoted in Simpson (1999), p. 151.
10. Comer (2000), p. 655.
11. Ibid., p. 656
12. Ibid., p. 588.
13. Maclean (2003), pp. 20–28
14. Ibid.
15. Comer, p. 658
16. Quoted in Simpson (1999), p.148
17. Bollinger, p. 28.
18. Licensing was profitable for the British Government and by the 1880s, 40 per cent of its revenue came from tax on alcohol. Quoted in Comer, p. 658.
19. Bollinger, p. 30.
20. Scott, p. 75.
21. Ibid., p. 120.
22. Stewart (2010), p. 78.
23. http://www.auntsfield.co.nz/historical-marlborough-vineyard.html; downloaded 12 January 2014.
24. Scott, p. 96.
25. Cooper (2008), p. 10.
26. Stewart (2010), p. 87.
27. Bragato, R. (1895), *Report on the Prospects of Viticulture in New Zealand.* Quoted in Thorpy.
28. Quoted in Scott, p. 92.
29. Thorpy, p. 52.
30. Scott, p. 96.
31. Ibid., p. 98.
32. Stewart (2010), p. 126.
33. Bollinger, p. 32.
34. Other electorates to vote for off-licence were Invercargill, Mataura, Oamaru, Roskill, Wellington South, and Wellington Suburbs.
35. Brien (2003), p.12
36. Bollinger, p. 32.
37. Rolfe (2000), p. 79
38. Brien, p. 22.
39. Malton and Cocker (no date). *Temperance and Prohibition, Period 1919–1928.*
40. Brien, p. 29.
41. Bollinger, p. 3.
42. Interview with James Browning Hay (b.1915), Christchurch, 17 October 2005.
43. Thomson (2012), p. 44.
44. Rolfe, p. 141.
45. Bollinger, p. 5.
46. Ibid., p. 147.
47. Stewart, p. 322.
48. Stewart, p. 209.
49. Thomson, p. 28.
50. Interview with Nick Nobilo, 10 Jan 2014.
51. Quoted in Thomson, J. (2012), pp. 58–60.
52. Rolfe, p. iv.
53. *A Brief History of New Zealand Wine Exporting* (2002).

Statistics

SUMMARY: NEW ZEALAND WINE (2004-2013)

	2004	2005	2006	2007	2008	2009	2010	2011	2012	2013
Number of Wineries	463	516	530	543	585	643	672	697	703	698
Number of Growers	589	818	866	1003	1060	1117	N/A	791	824	833
Producing Area (hectares)	18,112	21,002	22,616	25,355	29,310	31,964	33,428.0	33,400	35,337	35,182
Average Yield (tonnes per hectare)	9.1	6.9	8.2	8.1	9.7	8.9	8.0	9.8	7.6	9.7
Average Grape Price (NZ$ per tonne)	1,876	1,792	2,022	1,981	2,161	1,629	1,293	1,239	1,359	N/A
Tonnes Crushed	165,500	142,000	185,000	205,000	285,000	285,000	266,000	328,000	269,000	345,000
Total Production (millions of litres)	119.2	102.0	133.2	147.6	205.2	205.2	190.0	235.0	194.0	248.4
Domestic Sales of NZ Wine (millions of litres NZ Wine)	35.5	45.0	50.0	51.0	46.5	59.3	56.7	66.3	64.6	52.4[1]
Consumption per Capita NZ wine (litres NZ wine)	8.8	11.2	12.1	12.2	11.1	13.9	13.0	15.2	14.7	11.8[1]
Total sales of all wine (millions of litres)	79.7	81.7	86.0	91.8	87.4	92.7	92.1	93.9	91.9	93.3[1]
Consumption per capita all wines (litres)	19.6	19.8	20.6	21.7	20.8	21.5	21.1	21.3	20.9	21.1[1]
Export Volume (millions of litres)	31.1	51.4	57.8	76.0	88.6	112.6	142.0	154.7	178.9	169.6
Export Value (millions of NZ$ FOB)	302.6	434.9	512.4	698.3	797.8	991.7	1,041	1,094	1,177	1,211

[1] Estimate only

For updated statistics see www.nzwine.com

NEW ZEALAND WINEGROWERS MEMBERSHIP (2004–2013)

WINERIES BY CATEGORY[1]	2004	2005	2006	2007[1]	2008[2]	2009	2010	2011	2012	2013
Category 1	425	466	482	483	523	577	605	615	622	613
Category 2	34	44	42	51	56	60	61	73	71	75
Category 3	4	6	6	9	6	6	6	10	10	10
TOTAL	463	516	530	543	585	643	672	698	703	698

[1] Up to 2007: Category 1 – annual sales not exceeding 200,000 litres Category 2 – annual sales between 200,000 and 2,000,000 litres
Category 3 – annual sales exceeding 2,000,000 litres

[2] From 2008: Category 1 – annual sales not exceeding 200,000 litres Category 2 – annual sales between 200,000 and 4,000,000 litres
Category 3 – annual sales exceeding 4,000,000 litres

WINERIES BY REGION	2004	2005	2006	2007	2008	2009	2010	2011	2012	2013
Northland	8	10	10	11	14	14	14	15	16	13
Auckland	88	90	91	92	103	109	111	117	118	116
Waikato/Bay of Plenty	13	17	18	17	19	20	21	17	15	13
Gisborne	17	19	22	19	22	24	26	24	24	21
Hawke's Bay	58	62	66	67	71	79	85	91	84	77
Wairarapa	49	54	56	57	58	61	63	64	64	65
Nelson	24	29	29	28	32	34	36	38	36	38
Marlborough	84	101	106	104	109	130	137	142	148	152
Canterbury/Waipara	46	50	48	52	54	62	61	66	68	70
Central Otago	75	82	82	89	95	103	111	115	120	124
Other Areas	1	2	2	7	8	7	7	9	10	9
TOTAL	463	516	530	543	585	643	672	698	703	698

GRAPE GROWERS BY REGION	AUCK	WAIK	GISB	HB	WAIR	NELS	MARL	WAIP	CANT	OTAGO	TOTAL
2004	17	5	97	126	17	28	275	6	12	11	594
2005	18	7	108	168	33	40	415	7	12	17	825
2006	20	9	92	157	39	46	428	11	21	50	875
2007	25	4	100	186	25	58	530	12	4	63	1,007
2008	38	13	89	172	44	57	524	20	41	75	1,073
2009	44	11	87	171	48	62	568	22	38	77	1,128
2010	17	2	57	122	24	39	544	11	2	35	853
2011	9	2	54	103	24	38	551	6	2	35	824
2012	11	2	53	104	30	40	548	12	2	33	835

NEW ZEALAND PRODUCING VINEYARD AREA (2004–2013)

BY GRAPE VARIETY (HA)	2004	2005	2006	2007	2008	2009	2010	2011	2012	2013
Sauvignon Blanc	5,897	7,277	8,860	10,491	13,988	16,205	16,910	16,758	20,270	20,015
Pinot Noir	3,239	3,757	4,063	4,441	4,650	4,777	4,773	4,803	5,388	5,488
Chardonnay	3,617	3,804	3,779	3,918	3,881	3,911	3,865	3,823	3,229	3,202
Merlot	1,487	1,492	1,420	1,447	1,363	1,369	1,371	1,386	1,234	1,255
Riesling	666	811	853	868	917	979	986	993	770	787
Pinot Gris	381	489	762	1,146	1,383	1,501	1,763	1,725	2,485	2,403
Cabernet Sauvignon	687	614	531	524	516	517	519	519	305	300
Gewürztraminer	210	257	284	293	316	311	314	313	347	334
Syrah	183	211	214	257	278	293	297	299	387	408
Semillon	306	240	229	230	199	201	185	182	77	76
Cabernet Franc	213	180	164	168	166	163	161	161	119	119
Malbec	168	163	155	160	156	156	157	157	140	142
Muscat Varieties	136	139	140	139	135	135	125	125	48	49
Müller Thurgau	155	137	116	106	79	79	78	78	2	3
Pinotage	82	85	90	88	74	74	74	74	50	38
Chenin Blanc	72	58	59	50	50	50	47	47	21	26
Reichensteiner	61	59	61	66	72	72	72	72	14	14
Other & Unknown	552	1,229	836	963	1,087	1,171	1,731	1,885	449	525
TOTAL	18,112	21,002	22,616	25,355	29,310	31,964	33,428	33,400	35,335	35,182

BY REGION (HA)	2004	2005	2006	2007	2008	2009	2010	2011	2012	2013
Auckland/Northland	591	514	504	533	534	543	550	556	411	371
Waikato/Bay of Plenty	151	148	150	145	147	147	147	147	24	24
Gisborne	1,810	1,890	1,913	2,133	2,142	2,149	2,083	2,072	1,635	1,599
Hawke's Bay	3,873	4,249	4,346	4,665	4,899	4,921	4,947	4,993	5,030	4,846
Wairarapa	737	779	777	827	855	859	871	882	979	991
Marlborough	8,539	9,944	11,488	13,187	15,915	18,401	19,295	19,024	22,956	22,819
Nelson	548	646	695	782	794	813	842	861	1,011	1,095
Canterbury/Waipara	641	853	925	1,034	1,732	1,763	1,779	1,809	1,371	1,435
Central Otago	844	978	1,253	1,415	1,552	1,532	1,540	1,540	1,917	1,959
Other & Unknown	378	1,001	565	634	770	836	1,374	1,516	0	0
TOTAL	18,112	21,002	22,616	25,355	29,310	31,964	33,428	33,400	35,334	35,182

Source: New Zealand Winegrowers' Vineyard Surveys

NEW ZEALAND WINE TIMELINE

1769	Captain Cook arrives on *Endeavour*, returns on *Resolution* 1773, 1777
1814	Rev. Samuel Marsden visits New Zealand; conducts first church service
1819	Rev. Samuel Marsden returns to establish mission in Bay of Islands; plants grapes
1824	James Busby visits Bordeaux to study viticulture and winemaking
1825	James Busby emigrates to New South Wales; establishes farm in Hunter Valley
1831	Model farm established at Waimate Mission, Bay of Islands
1832	James Busby visits wine growing regions in Jerez, Rhône, Burgundy
1833	James Busby arrives as British Resident; plants grapes at Waitangi
1835	Charles Darwin visits on the *Beagle*
1838	Bishop Pompallier arrives from France to set up Roman Catholic missions
1840	Treaty of Waitangi
	French Explorer Dumont d'Urville tastes Busby's wine
	New Zealand Company settlers arrive in Wellington
	French settlers arrive in Akaroa
1841	Distillation Prohibition Ordinance; distillation only for medicinal purposes
1842	Licensing Ordinance prevents the sale of liquor except by licensed people
1843	German winemaking settlers arrive in Nelson, but soon leave
1851	Marist brothers plant grapes at Hawke's Bay mission
1863	Charles Levet plants vines at Kaipara
1864	Jean Desire Feraud plants grapes at Monte Cristo, near Clyde, Central Otago
1868	Distillation Act controls distilling
1870s	Heinrich Breidecker makes wine in Hokianga

1873	David Herd plants grapes at Auntsfield, Marlborough.
1874	Excise Duties Act force distilleries to close
1880	Joseph Soler from Wanganui wins six awards in Melbourne
1881	Licensing Act establishing licensing districts and committees
1883	William Beetham plants vines in Wairarapa
1890	Government grants winemakers right to operate their own stills to fortify wine
1893	Electoral Act 1893 gives all women in New Zealand the right to vote
1894	Clutha electorate goes dry
1895	Visit by Romeo Bragato; identifies phylloxera pest
1896	Dalmatian immigrants plant first vines
1897	167 hectares of vines nationally
1899	Further outbreaks of phylloxera prompt calls for Bragato to return
1902	Bragato returns as Government Viticulturist.
	Mataura and Ashburton go dry
	Assid Abraham (A.A.) Corban plants vineyard in Henderson
1905	Grey Lynn, Oamaru, Invercargill go dry
	Anthony Vidal, nephew of Joseph Soler, plants vines in Hawke's Bay
1906	Joseph Soler wins five gold medals at NZ International Exhibition
	Bragato publishes *Viticulture in New Zealand* handbook
1908	Masterton and Eden go dry
1909	Bragato leaves New Zealand. Now 269 hectares of vines nationally
1910	Growth of prohibition as 12 of 76 electorates go dry =16%
1911	Referendum on prohibition defeated; Drinking age raised to 21 years
1914	First World War begins

1917	Six o'clock closing introduced under Sale of Liquor Restriction Act
1919	National Referendum on prohibition defeated by votes of returning servicemen
	December 1919, second referendum fails by 2000 votes
1920s	Prohibition pressure starts to retreat
1927	Tom McDonald purchases Taradale vineyard from his employer
1930s	Waves of Dalmatian immigrants settle in West Auckland
1934	Ivan Yukich founds Montana wines
1935	Labour Government introduces import licences and duties on imported wines
1939	Second World War begins
1942	Influx of American servicemen; demand for wine grows steadily
1943	Nikola Nobilo plants vines in Huapai
1944	Tom McDonald sells to Ballins Breweries; rename winery as McDonald Wines
1945	World War Two ends; many returning servicemen have developed taste for wine
1947	Nikola and Vidosava Delegat plant vines near Henderson
1948	Wine resellers licence allows growers to set up wine retail outlets
1949	Frank Berrysmith appointed government viticulturist
1951	George Mazuran begins crusade to change liquor licensing laws
1955	Government reduces minimum quantities of table wine to 750 ml bottle
1956	Mate Selak releases Champelle, first New Zealand *méthode traditionnelle*
1958	Import licences for wine, spirits halved
1960	First licensed restaurants
1961	Montana wines release *Pearl*, in bulb-shaped bottle with screwcap; Cold Duck

1961	George Fistonich establishes Villa Maria
1962	McDonalds Wines under McWilliams; produces Cresta Doré (white), Bakano (red); Marque Vue (sparkling) labels
	Alex Corban pioneers sparkling Premiere Cuvée using Charmat process; Riverlea Riesling (Muller-Thurgau)
	Frank Berrysmith imports Muller-Thurgau (Riesling Sylvaner) from Germany
1963	Penfolds buy winery in Henderson
1965	McWilliams Cabernet Sauvignon becomes New Zealand's most sought after wine
1968	Bill Irwin establishes Matawhero, Gisborne
1969	Cooks New Zealand Wine Company set up by Auckland businessmen
1970s	Growth in Gisborne; contract growers plant Muller-Thurgau and hybrids
	Chardonnay Mendoza Clone imported along with UC Davis Chardonnay clones 4, 5
	Malcolm Abel confiscates Pinot Noir cutting at Auckland Airport customs.
	Ross Spence makes experimental Sauvignon Blanc
1973	Montana buys land in Marlborough to grow grapes; plants Sauvignon Blanc
	Peter Hubscher becomes Montana's chief winemaker; Montana Gisborne Chardonnay is produced
	Seagram acquires 40% of Montana
1974	Te Mata Estate winery is acquired by John Buck and partner
	Matua Valley established by Ross and Bill Spence
1976	NZ Wine Institute set up to represent and promote the wine industry.
	Villa Maria buys Vidal
1980s	Tax incentives to plant grapes result in overproduction and fierce price wars

1980	Government makes it illegal to add water during winemaking
1981	Montana launches Lindauer Méthode Traditionnelle
	Dr Richard Smart appointed viticultural scientist at Ruakura
	Danny Schuster and David Jackson publish *Grape Growing and Winemaking*
1982	First London tasting of wines at New Zealand House
	St Helena Pinot Noir wins medal
1984	Increase of sales taxes on Sherries, Ports by 54% imposes hardships on growers
1985	Wine surplus causes Cooks McWilliams (now merged) to slash prices to sell stock
	New Zealand Society for Viticulture and Oenology (NZSVO) establised to promote the dissemination of technical information in viticulture and oenology
	Cloudy Bay releases first Sauvignon Blanc
1986	Government vine-pull scheme to stabilise industry. Growers paid to uproot vines
	Some growers replant with fashionable varietals like Sauvignon Blanc
	Hunter's Fumé Blanc 1985 voted top wine by the public at London Wine Show
1987	NZ has first stand at London Wine Trade Fair
	Seagram exit Montana Peter Masfen buys shares and brings Montana under control of Corporate Investments Ltd
	Montana buys Church Road
1989	Sale of Liquor Act allows wine to be sold in supermarkets; easier to apply for on-premise licences
1990	Montana Sauvignon Blanc wins trophy at London International Wine Challenge
1991	Wine Guild established to promote export of New Zealand wines
1993	Nobilo exports over 1 million litres of White Cloud for bottling in Sweden

2000	Montana buys Corbans
	BRL Hardy buys Nobilo's, who are then taken over by Constellation Wine Group
	Cloudy Bay becomes 100% owned by LVMH
2001	New Zealand Screwcap Wine Seal Initiative
	Montana sold to Allied Domecq. Matua sold to Fosters (later Treasury Wine Estates)
2003	Wine Act
	New Zealand Winegrowers formed to combine NZ Wine Institute and NZ Grape Growers Council
2006	New Zealand's Geographical Indications (Wine and Spirits) Registration Act 2006 to define and protect names of winegrowing regions and localities.
	Growth in Awatere led by Yealands
	Grove Mill (NZ Wine Company) becomes the first certified carbon neutral (carboNZero) winery in the world
2007	Sustainability policy with goal that by 2012 all New Zealand wines will have audited environmental programmes
2008	Largest vintage to date with 285,000 tonnes crushed; concerns about over production
	Impact of Global Financial Crisis (GFC) starts to affect international sales
2010	Exports exceed $1 billion with Sauvignon Blanc the key driver
2011	Organic winegrowing policy with goal of 20% certified organic vineyards by 2020
2012	Consolidation of winegrowers as industry focuses more on exports
2014	Focus on export markets and building brand identity of New Zealand wines

BIBLIOGRAPHY

Auntsfield Estate: Marlborough's first and oldest vineyard (no date), downloaded 12 January 2014: http://www.auntsfield.co.nz/historical-marlborough-vineyard.html

Bird, David (2005) *Understanding Wine Technology*, DBQA Newark, UK

Bollinger, Conrad (1967) *Grog's Own Country: the Story of Liquor Licensing in New Zealand*, Minerva, Auckland, NZ

Brien, B. (2003). *100 Years of Hospitality in New Zealand 1902-2002*. Wellington: Wellington Museums Trust

Busby, James (1842) *Catalogue of vines in the Botanic Garden, Sydney, introduced into the colony of New South Wales in the year 1832*, W.J. Row, Government Printer, Sydney, Australia

Cocker, J. Malton Murray, J., (eds) (1930) *Temperance and Prohibition in New Zealand: compiled and issued under the auspices of New Zealand Alliance for the Abolition of the Liquor Traffic*, Epworth Press, London, UK

Comer, J. (2000) *Distilled Beverages. In: Kiple, K. and Ornelas, K. (eds). The Cambridge World History of Food Vol 1.* (pp. 653-663). Cambridge: Cambridge University Press

Cooper, Michael (2008) *Wine Atlas of New Zealand*, 2nd ed., Hodder Moa, Auckland, NZ

Cooper, Michael (2011) *Buyer's Guide to New Zealand Wines 2012*, Hodder Moa, Auckland, NZ

Courtney, Caroline (2003) *Wine in New Zealand*, Random House, Auckland, NZ

The Dictionary of New Zealand Biography, Vol. 1. (1990) W.H. Oliver, ed. Allen & Unwin, Wellington, NZ

Hay, Celia (2006) *How to Grow your Hospitality Business*, 2nd ed., H and H Publishing, Christchurch, NZ

Hanni, Tim (2013) *Why you like the wines you like*, HanniCo LLC, Napa, USA

Jackson, David (2004) *The Wine Drinker's Guide to the Vineyard*, Dunmore Press, Palmerston North, NZ

Johnson, Hugh and Robinson, Jancis (2013) *World Atlas of Wine*, 7th ed., Mitchell Beazley, London

Judd, Kevin (2009) *The Landscape of New Zealand Wine*, Craig Potton Publishing, Nelson, NZ

MacLean, C. (2003). *Scotch Whisky: a liquid history.* London: Cassell

New Zealand Organic Market Report (2012), downloaded 15 January 2014: http://www.oanz.org/casestudies/OANZ-full%20report%202012.pdf

Paton, Clive, *The Gumboot Clone*, downloaded 19 January 2014: http://www.atarangi.co.nz/the-gumboot-clone.html

Robinson, Jancis (ed) (2006) *Oxford Companion to Wine*, 3rd ed., Oxford University Press, UK

Robinson, Jancis et al., (2012) *Wine Grapes*, Harper Collins, NY, USA

Rolfe, J. (2000) *In the Club: a history of the chartered club movement in New Zealand.* Auckland: Celebrity Books

Saker, John (2010) *Pinot Noir*, Random House, Auckland, NZ

Scott, Dick (2002) *Pioneers of New Zealand Wine*, Reed/Southern Cross Books, Auckland, NZ

Schuster, Danny, Jackson, David and Tipples, Rupert (2002) *Canterbury Grapes and Wines 1840-2002*, Shoal Bay Press, Christchurch, NZ.

Simpson, Tony (1999) *A Distant Feast.* Auckland: Godwit/Random

Spang, Rebecca (2000) *The Invention of the Restaurant: Paris and Modern Gastronomic Culture.* Cambridge: Harvard University Press

Stewart, Keith (2010) *Chancers and Visionaries: A History of Wine in New Zealand*, Godwit, Auckland, NZ

Thomson, Joelle (2012) *The Wild Bunch*, New Holland Publishers, Auckland, NZ

Thorpy, Frank (1983) *Wine in New Zealand*, Collins, Auckland, NZ

Trubek, A. B. (2000) *Haute Cuisine: how the French invented the culinary profession.* Philadelphia: University of Pennsylvania Press.

Tyack, Kerry (2012) *The Winemaker: George Fistonich and the Villa Maria Story*, Random House, Auckland, NZ

Wine and Spirit Trust (2012) *Wine and Spirits: Understanding Style and Quality*, 2nd ed., WSET, London, UK

Websites:

www.nzwine.com

www.jancisrobinson.com

Encyclopaedia of New Zealand: story of wine: http://www.teara.govt.nz/en/wine/page-1

www.myvinotype.com

INDEX